ALSO BY LISA MARGONELLI

Oil on the Brain: Petroleum's Long, Strange Trip to Your Tank

UNDERBUG

UNDERBUG

AN OBSESSIVE
TALE OF
TERMITES
AND TECHNOLOGY

Lisa Margonelli

Illustrations by Thomas Shahan

SCIENTIFIC AMERICAN | FARRAR, STRAUS AND GIROUX

New York

Scientific American / Farrar, Straus and Giroux
175 Varick Street, New York 10014

Copyright © 2018 by Lisa Margonelli
Illustrations copyright © 2018 by Thomas Shahan
All rights reserved
Printed in the United States of America
First edition, 2018

Portions of chapters 18 and 19 originally appeared,
in slightly different form, in *Scientific American*.

Library of Congress Cataloging-in-Publication Data
Names: Margonelli, Lisa, author.
Title: Underbug : an obsessive tale of termites and technology / Lisa
    Margonelli.
Description: First edition. | New York : Scientific American / Farrar, Straus
    and Giroux, 2018. | Includes bibliographical references and index.
Identifiers: LCCN 2017059906 | ISBN 9780374282073 (hardcover)
Subjects: LCSH: Termites.
Classification: LCC QL529 .M374 2018 | DDC 595.7/36—dc23
LC record available at https://lccn.loc.gov/2017059906

Designed by Richard Oriolo

Our books may be purchased in bulk for promotional, educational, or
business use. Please contact your local bookseller or the Macmillan Corporate
and Premium Sales Department at 1-800-221-7945, extension 5442,
or by e-mail at MacmillanSpecialMarkets@macmillan.com.

www.fsgbooks.com • books.scientificamerican.com
www.twitter.com/fsgbooks • www.facebook.com/fsgbooks

Scientific American is a registered trademark of Nature America, Inc.

1   3   5   7   9   10   8   6   4   2

FOR MINA

-4|

When we come under the spell of the deeper domain of [technology], its economic character and even its power aspect fascinate us less than its playful side. . . . This playful feature manifests itself more clearly in small things than in the gigantic works of our world. The crude observer can only be impressed by large quantities—chiefly when they are in motion—and yet there are as many organs in a fly as in a leviathan.

—ERNST JÜNGER, *THE GLASS BEES*

CONTENTS

PART 1

1. A Termite Safari   3

PART 11

2. Riddles in the Dirt   15
3. An Inconvenient Insect   26
4. Into the Mound   33
5. Complexity Is the Essence   47
6. Because They Are So Sweet   53
7. A Black Box with Six Legs   60
8. Waiting for Carnot   73

PART 111

9. The Second Termite Safari   83
10. Life in the Firehose   96
11. Jazz in the Metagenome   107
12. Burning Very Slowly   117
13. Restless Streams   127

PART IV

14. Crossing the Abstraction Barrier   135
15. Influential Individuals   148
16. The Robot Apocalypse   160

PART V

17. Darwin's Termites   173
18. The Soul of the Soil   186
19. The Math of Fairy Circles   195
20. The Soul of the Cell   207
21. Empathy and the Drone   218
22. White Ants   229

PART VI

23. *Them* and Us   247

NOTES   257
ACKNOWLEDGMENTS   289
INDEX   291

PART 1

ONE

A TERMITE SAFARI

ARIZONA

IN JULY 2008, I RENTED a small yellow car in Tucson, Arizona, and drove it south toward Tombstone. There were three passengers in my car, and I was following a white van with government plates carrying nine more. Between these two vehicles we had eleven microbial geneticists from six countries with nearly three hundred years of collective education. We also had five hundred plastic bags, a thermos of dry ice, and three hundred fifty cryogenic vials, each the size and shape of a pencil stub. We had two days to get ten thousand termites.

I was there because I'd received an email titled "Termite Safari?" from Phil Hugenholtz, a researcher at the U.S. Department of Energy Joint Genome Institute (JGI). Earlier that year I'd done a story for *The*

Atlantic about Phil and thirty-eight other scientists who sequenced a million genes from the microbes found in the guts of *Nasutitermes corniger*—termites they'd found in trees in Costa Rica. Because termites are famously good at eating wood, the genes in their guts were attractive to government labs trying to turn wood and grass into fuel: "grassoline." The white van and the geneticists all belonged to JGI.

As I drove, the price of gasoline was more than four dollars a gallon, the highest it had ever been. The country, maybe the world, was spinning into a financial crisis, and a vicious presidential race was on. In those days, gasoline was always on my mind because I'd written a book about oil. When I received Phil's email, I had been writing about the problems of petroleum for seven years, and I was staring at twenty open PDFs about energy on my computer's desktop. I was at the end of my rope, both as an oil-consuming citizen and as a human being. Oil's troubles are systemic, dating back at least a hundred years—and they're made much nastier by modern politics. Termite Safari? Sure! Anything was better than sitting at my desk and thinking about oil.

Microbial geneticists are young and goofy: they sincerely believe in the scientific method and they blink in the sun. Phil, who led the lab with a jangly self-deprecating pride, called them "gene jockeys." They lived in statistics and databases and thought in codes, both computer and genetic. Their organization was so antihierarchical that I sometimes had to tell them to stop bickering. "Stop it or I'll turn this car right around," I'd say.

Left on their own, the geneticists would never find ten thousand termites. They might not even find one hundred. North American termites are cryptic in the extreme: they live in tunnels underground and inside wood. So Rudi Scheffrahn, one of the country's leading termite experts, sat in the passenger seat next to me, charting a course to the bugs.

He wore bifocals over his sunglasses while he scoured the last survey of termites in the area, published in 1934, around the time of the building of the Hoover Dam. Funded by a sugar refiner, along with lumber, railway, and electric companies, the survey hired entomologists to figure out which bugs might try to eat the electrical poles, bridges, and railway trestles that industry was erecting across California, Nevada, Arizona, and Mexico.

When Rudi sensed termites in the landscape, his sunburned knees began to twitch, which jiggled his head, sending little triangles of reflected light from his two pairs of glasses ricocheting around the rental car. It felt like a party.

When there were no termites, he seemed depressed. We drove past a lake. "Probably a nice place," he said, in a flat, declining tone. "But there's no termites there."

Rudi is a termite chauvinist. He talked world records. Oldest insect: termite queens can live to be twenty-five years old, maybe more, no one knows. Fastest animal: some soldier termites slam their mandibles shut at 120 miles an hour, faster than a cheetah runs. Biggest terrestrial structures: mounds in Africa, Asia, and Australia can be up to thirty feet high. And about termites' reputation as pests? Only twenty-eight out of twenty-eight hundred species are invasive pests, and Rudi believes that the first destructive drywood termites traveled from Peru on Spanish ships in the 1500s. Not their fault.

The geneticists in the back of the car laughed politely. They were not here to appreciate termites; they were here to appropriate them. Termite guts are a molecular treasure chest: 90 percent of the organisms in them are found nowhere else on Earth. The purpose of this trip was to gather the termites and their microbial mush in the vials and jam them into the dry ice before the genetic information they held deteriorated. The geneticists didn't just want the microbes' DNA, they also wanted the molecules of RNA, which could tell them which parts of the genetic code were in use at the precise moment the termites took their tumble into the thermos. Perhaps by seeing exactly how termites break down wood, we'd be able to do it, too.

If this trip succeeded in capturing a few good molecular moments, humans might eventually be able to power our cars without worsening climate change, or make fuel without drilling in national parks or causing oil spills. This trip could change the world—or at least the lives of the scientists in the cars.

We barreled down a long hill toward a scrub basin when Rudi's legs began to twitch. The reflections from his glasses whizzed around the interior of the car. "I look out here and I just see billions of termites."

JUST BEFORE I got interested in termites, they lost the identity—the distinctive order Isoptera—that they'd had for the last 175 years. In 2007 a paper was published called "Death of an Order: A Comprehensive Molecular Phylogenetic Study Confirms That Termites Are Eusocial Cockroaches." And just like that, termites became homeless and nameless, demoted to gregarious cockroaches, very distant cousins of the praying mantis. Maybe more important, they were suddenly defined not by what they looked like, but by their genes. Their relationship to the most despised of bugs wasn't news: it was first explored in the 1920s, and became obvious with DNA sequencing by 2000 or so, but it wasn't considered a fact until 2007.

Here's a possible story of how termites came to be—and why we were chasing them across Arizona. Once upon a time, cockroaches were solitary scavengers that ate fruit, rotten leaves, fungi, and bird droppings. They developed their eggs in beautiful bean-like sacs, shot them out their backsides at high speeds, and walked away, leaving their young to fend for themselves.

Sometime between 250 million and 155 million years ago, termites evolved from the chassis of these ancient roaches when some of them ate microbes that were able to digest wood. Wood was plentiful, but to survive on it the bugs needed to have those microbes handy at all times. The problem was that they regularly molted their intestines, which cleaned the microbes right out. Our evolving cockroaches started to exchange what entomologists politely call "woodshake"—a slurry of feces, microbes, and wood chips—among themselves, mouth to mouth and mouth to butt. After they pooled their digestion, it was a quick trip to constant communal living. One for all and all for one: Termites forever!

Termites' twin innovations—wood-eating microbes and

eusociality—are each worthy of an invertebrate Nobel Prize, but millions of years of evolution refined them still further. While the dinosaurs stomped around outside, these proto-termites evolved inside their dark logs, improving their togetherness by communicating through chemicals, sound, and touch. The queen, who produced all of the eggs for a colony, secreted chemicals that kept most of her children from reproductively maturing. Most of them were workers, gathering food, feeding the queen and the young, and doing tasks around the nest, but some differentiated into what we call soldiers, with ingenious strategies to better sacrifice themselves for the survival of the colony: nozzles filled with noxious chemicals on their heads, slashing mandibles, or big wedgy heads they could use to seal off passageways from invaders. Without the need to reproduce, or to venture far aboveground, both worker and soldier termites lost things they didn't need: eyes, wings, and big, tough exoskeletons.

By the time Pangaea split up and an ocean grew in between Brazil and Africa, termites were well positioned to evolve in ever weirder ways on different continents, in a big belt stretching around Earth at the equator and extending halfway to the North and South Poles.

FROM TOMBSTONE WE drove south to Bisbee, Arizona, and then toward the border with Mexico, stopping whenever Rudi felt termites. I finally found some on my own in the Coronado National Forest, when I wandered far from the road in the higher-altitude desert there. The light was pinkish and flat and the air smelled lightly of sage. Rudi said to look for termites under rocks because they'd be orienting themselves thermally. So I threw over a flat rock and got down on the ground. To capture the termites I had an aspirator—a homemade contraption consisting of a test tube, a tiny air filter, and some tubing—that allowed me to suck up specimens without getting a mouthful of termite.

Under the first rock I saw only beetles, grubs, white spiders, millipedes, and ants. When I looked up, I saw a white Homeland Security truck with a cage on the back, roaming the land like a hu aspirator. In a gully were old sneakers, flattened Mylar hydration packs, and other evidence that some person or persons had come through from the border with Mexico. I hated to think of someone hiding, dehydrated, in this no-man's-land. My phone chuckled intermittently, delivering texts that said I'd crossed into Mexican telephonic space and different rates would apply. The scrub here was barer than in other places we'd stopped. I wondered if it was recently vacated or secretly occupied. I didn't so much feel eyes upon me as I felt an uncanny tingle.

The landscape made me feel dumb and clumsy. If I had to be dumb, I wanted to be systematic about it. I began turning over every stick, rock, and rotten cactus in my path. Termites had chewed the underside of most sticks and left behind eerie little wooden galleries—ghost towns with tiny, sticky dust webs fluttering like tattered lace curtains.

I had almost forgotten I was searching when I lifted a rock and saw a glint of glossy exoskeleton flowing into some little tunnels. I dropped to my knees and began sucking on the aspirator, a disgusting process that stimulated saliva production and made me dizzy. Two minutes later there were no more termites on the ground and I had about twenty-five in the test tube attached to the aspirator.

After the long hunt, my pale termites were disappointing. When I separated one from the clutch, it was less substantial than a baby's fingernail clipping. Doddering around blindly, it waved the flimsy antennae on its bulbous head. In its stubby, translucent body I could almost see its coiled guts—and presumably whatever it had eaten for lunch. Ants have snazzy bodies with three sections, highlighted by narrow waists, like pinup models', between the segments. Termites, which are no relation to ants or bees, have round, eyeless heads, thick necks, and teardrop-shaped bodies. And they long ago lost the cockroaches' repulsive dignity, gnarly size, and gleaming chitinous armor. I put the termite back in the test tube.

What had I just sucked up? My little gang of twenty-five was incapable of doing much of anything. Without a colony, they had nowhere to bring food to, and thus no reason to forage. Without a crowd of soldiers, they couldn't defend themselves. Without a queen, they couldn't reproduce. Twenty-five termites are insignificant in the scheme of life and death and reproduction. Meaningless. What's more, they were clinging to one another, making an icky beige rope of termite heads, bodies, and legs reminiscent of the game Barrel of Monkeys. In the miniature scrum I couldn't even see a single termite—they looked like a clot, not a group of individuals. Or perhaps I'd found a single individual who happened to have twenty-five selves.

I'd stumbled into one of the big questions termites pose, which is, roughly, what is "one" termite? Is it one individual termite? Is it one termite with its symbiotic gut microbes, an entity that can eat wood but cannot reproduce on its own? Or is it a colony, a whole living, breathing structure, occupied by a few million related individuals and a gazillion symbionts who collectively constitute "one"?

The issue of One is profound in every direction, with evolutionary, ecological, and existential implications. The word used to describe this many-in-one phenomenon these days is "superorganism," but at the time I didn't know much about that concept.

By the end of that day I still couldn't see termites the way Rudi could, but I had a basic idea that the fewer I saw, the more termites there might be. Where I had thought of landscapes as the product of growth, on that afternoon they inverted to become the opposite: the remainders left behind by the forces of persistent and massive chewing. The sky was no longer the sky, but the blue stuff that is visible after the screening brush and cacti have been eaten away. Termites have made the world by unmaking parts of it. They are the architects of negative space. The engineers of not.

NOBODY LOVES TERMITES, even though other social insects such as ants and bees are admired for their organization, thrift, and industry. Parents dress their children in bee costumes. Ants star in movies and video games. But termites are never more than crude cartoons on the side of exterminators' vans. Termite studies are likewise a backwater, funded mostly by government agencies and companies with names like Terminix. Between 2000 and 2013, 6,373 papers about termites were published; 49 percent were about how to kill them.

Every story about termites mentions that they consume somewhere between $1.5 and $20 billion in U.S. property every year. Termites' offense is often described as the eating of "private" property, which makes them sound like anticapitalist anarchists. While termites are truly subversive, it's fair to point out that they'll eat anything pulpy. They find money itself to be very tasty. In 2011 they broke into an Indian bank and ate $220,000 in banknotes. In 2013 they ate $65,000 that a woman in Guangdong had wrapped in plastic and hidden in a wooden drawer.

Another statistic seems relevant: termites outweigh us ten to one. For every 132-pound human you, according to the termite expert David Bignell, there are 1,320 pounds of *them*.

We may live in our own self-titled epoch—the Anthropocene—but termites actually run the dirt. They are our underappreciated underlords, key players in a vast planetary conspiracy of disassembly and decay. If termites, ants, and bees were to go on strike, the tropics' pyramid of interdependence would collapse into infertility, the world's most important rivers would silt up, and the oceans would become toxic. Game Over.

BY THE END of the trip we had eight thousand termites in Tupperware tubs and plastic bags, but they needed to be labeled and killed before going to California to be sequenced. Phil borrowed a lab from a colleague at the University of Arizona, and we all set to work hustling the termites into small plastic vials. Each vial was given a letter and some numbers in accord with a complicated system of geographic location, food, and species. Anna Engelbrektson, the lab's tech, sat cross-legged on a lab counter in a bubble of Sharpie-scent, labeling at a high rate of speed while telling the postdocs what to do. Then she tossed containers into the dry ice and that was that. Frozen in the thermos, the termites were on their way to immortality: a collection of genetic code sitting in some database on a server somewhere, intellectual property, a sequence of nucleotides that might solve a wicked problem someday.

That night in the lab we sat on the border between natural history and an unnatural future. We weren't alone: all over the world, scientists are trying to find biology's underlying rules and put them to use. They're doing it with genes, behaviors, metabolisms, and ecosystems. They're seeing nature in new ways, and at the same time they're trying to reinvent it and put it to work for us.

We started this trip with Rudi retracing the termite survey from 1934, just before the Hoover Dam altered the course of the Colorado River, put 247 square miles of land underwater, and massively relocated both water and power throughout the West. That dam now sits like a crown between two mountains—an awe-inspiring machine-age masterpiece, harnessing the big forces of nature to tie the globe together in a web of power lines, pipelines, refineries, satellites, and fiber optic cables.

But that was all the old future. In the next future, we will harness nature's tiniest life-forms—microbes and insects—both their systems of organization and control, and their genes and chemical capabilities. This fits with our paradoxical desire to have a lighter footprint on the Earth while having greater control over its processes.

At the core of this project is the provocative dream of changing biology into a predictive science, much the way physics started as the observation of phenomena like gravity and then became the science of making plans for the atom bomb. Will there be termite bombs? Whether the answer to that is yes or no, our grandchildren will laugh at the innocence of our question.

Termites, I came to understand, are the poster bug for the twenty-first century—a little guide to really big ideas.

I PUT MY HEAD down on a table to nap. I had no idea where my interest in termites would take me over the next eight years. I didn't realize I would become obsessed and spend many hours reading research papers. Or that I would cold-call scientists until I had a collection of friendly roboticists, computer scientists, physiologists, ecologists, synthetic biologists, physicists, geneticists, and mathematical biologists who let me watch them work in places as far-flung as California, Namibia, Massachusetts, and Australia. Nor did I have any premonition of the curious things I'd meet along the way: the northerly spires, the singing queen bee, the charismatic termite, the paradoxical creature with the mixed-up hairs, and the white ants.

I awoke a little before eleven, having dreamed of moving spots. With the sorting and freezing almost done, Anna was exhorting the postdocs to start cleaning. As I got up to help, I noticed that all of us except Rudi were wearing stonewashed jeans. They were in all the stores then, so there was nothing obviously extraordinary about this. After my sun-scorched days in termite land, though, the pants looked like exoskeletons, protecting us from thorns and sun. Reproductively, they identified us to fellow members of the species "mall shopper," signaling sexy industriousness to potential mates.

As I started sweeping, I recalled that the thing about stonewashed jeans is that there are no stones, only genes. In those days, one of the biggest markets for genetically modified proteins was enzymes that could be painted onto blue jeans to make them look faded and worn—replacing an older process of battering them with stones. On balance, the enzyme process was better for the planet, but I always get hung up on how funny it is that we have devoted decades of scientific study to make new pants look like they've been out digging ditches.

Borrowing genes from the environment—whether for pants or to turn wood into fuel—is a very termitey thing to do. Were we unwittingly becoming somewhat like our quarries? I had hardly spent any time with termites and already my fellow humans were looking a little strange.

PART 11

TWO

RIDDLES
IN THE DIRT

NAMIBIA

IT TAKES ALMOST A DAY to fly from the United States to Windhoek, the capital of Namibia. First I went to Germany and then on to Johannesburg and from there a short hop westward to Namibia, which is a wedge of sun-blasted land between South Africa and the Atlantic Ocean, capped by Angola to the north. As we prepared to land, the ground appeared to be a finely arranged carpet of scrub, like a Persian rug with a pattern I couldn't quite take in. Once on the ground I got a rental car and began driving north toward the town of Otji-warango, on assignment from *National Geographic* to write about J. Scott Turner, an American biologist who has spent decades studying how and why termites build their mounds. Along the way Scott's investigations

had taken a turn into computer simulations and speculative collaborations with an engineer who sees termites' ability to build mounds as
an inspiration for constructing shelters on Mars.

Namibia only has a few paved roads, and this one, the B1, was perfectly flat, nearly empty, and as straight as if it had been drawn on a
map with a ruler. Once Windhoek's city limits were behind me, I could
see the land that had looked like a carpet from above, but now I perceived it in horizontal layers: first brownish dirt, then shimmering
green acacia bushes, and standing above the scrub, symmetrical trees
that resembled giant broccoli stalks. As I drove northward the dirt became a garish matte orange and then lurid salmon pink. Other times
it was dark brown or beige.

After an hour or so a spire of dirt appeared beside the road. About
fifteen feet high, the dirt appeared to have been tossed up from below
the ground's surface, the way Bugs Bunny tunneled before he took a
wrong turn at Albuquerque. I was excited to see it: I knew from reading Scott's papers that this was a *Macrotermes* mound, home to several
million termites. I also knew that below the mound sat a large fungus
that the termites tended. What I did not expect was that the top of the
mound would lean so.

More mounds appeared beside the road. Their shape was clearly
not accidental. If a fist made of dirt punched through the surface of the
ground from below and then extended one finger to point at the sky,
it would account for the mounds' aggressive posture and distinctively
wide base and narrow spire. Then there were dozens of them. And then
many dozens. Their colors shifted with the dirt around them: Cream-
sicle orange, then bone, then salmon, then a solid homespun brown.
All of these dirt fingers appeared to be pointing north, as though
directing me to my destination.

Soon I could see nothing but termite mounds to the horizon
on both sides. I had been traveling for many hours, and I began
to play with them in my mind. Were they a troop of thousands of
mournful monks in robes? Or were they comic? Maybe they were a
Busby Berkeley chorus line, each dancer holding her straw hat aloft

at the same inviting angle in the direction I was driving: Come with us!

I had no time to stop and investigate because I needed to meet Scott in Otjiwarango before dusk, when—yellow signs warned—kudu and warthogs would begin jumping into the roadway. As I drove I mused about termites, the puzzling straightness of this road, and the odd arrangement of the termite mounds beside it.

TERMITE COLONIES BEGIN theatrically on rainy evenings. Small holes open in the sides of existing termite homes and largish, winged termites emerge, shake out their sticky wings, and fly. In Northern California termites of the genus *Reticulitermes* suddenly appear on the sides of buildings they inhabit. In South America *Nasutitermes* shower down from nests in the trees. In New Orleans, Formosan termites, of the genus *Coptotermes*, burp from colonies in the ground and take to the air in swarms so dense they show up on weather radar. In Namibia these giant *Macrotermes* mounds seem to spring a leak, spilling froths of winged termites down their sides.

In the mound, most of the termites are eyeless and wingless, but the fertile termites who leave the mound on this night have eyes and what at first appears to be one single translucent teardrop-shaped wing, starting at their heads and ending, rounded, at their hinds. When they are ready to fly, this single wing, still soft and moist, fans out into four. Called "alates," these termites are like fragile balsa wood glider planes: just sturdy enough to cruise briefly before crash-landing their payloads of genes.

Alates are scrumptiously fatty, and reportedly have a nutty flavor, so what starts as a confetti shower of gametes turns into a scrum of birds, lizards, aardwolves, and sometimes humans trying to gobble them up, with the result that hardly any survive this nuptial flight. It's possible that catching and eating these termites gave our australopithecine ancestors a booster shot of fat, proteins, and micronutrients that helped to feed their growing brains, leading eventually to our current human situation. This strange fact—that termites themselves may be partly responsible for the brains with which we try to study them—is typical of the weird dual vision of studying termites. They are both utterly alien and strangely familiar.

One of the greatest storytellers of the termite is the South African writer Eugène Marais, who spent many years peering into their mounds. Here is how he describes what happens next in the nuptial flight, when a single female has just landed on a bit of damp ground:

> The first thing she does is to discard her wings. This she succeeds in doing by a lightning-like movement—so fast that we cannot follow it with the eye. One moment we see her with her wings intact, the next moment she steps away, and her four wings are lying on the grass—she is much, much quicker than a woman who discards her evening gown and hangs it over a chair. It took months for her wings to grow. For years perhaps she has lived in subterranean darkness, in preparation for this one moment. For a period of three seconds, for a distance of perhaps three yards, she enjoyed the exquisite thrill of flight and with that the object of a great preparation has been fulfilled and the fairy-like wings are flung aside like a worn-out garment. . . .
>
> She comes to rest on her fore-feet and lifts three-quarters of the hinder part of her body into the air, and she remains stationary in this position, as still as if she were merely the statue of a termite.

If you want to learn about ants, you can read books by great scientists like E. O. Wilson, Bert Hölldobler, Deborah Gordon, and the photographer Mark Moffett. My local bookstore has so many books devoted to bee economics, bee democracy, bee dancing, and more, that you could mistake it for a self-help department for people hoping to take up life as bees. But when I went hunting for books on termites, there were academic texts and there was Marais's *The Soul of the White Ant,* originally published in English in 1937. Everything about the book is ridiculous, starting with its name and going on from there. What sort of writer has women and termites alike throwing off their evening gowns with exquisite haste? His tale of the termite mound is part close observation, part poetic riddle, and part thumbnail guide to the universe, but it's not exactly scientific. Still, nobody since has gotten further into imagining the thoughts of the mound than Marais, making his book an invaluable document—of our minds more than theirs.

Marais was a journalist, a poet, and a lawyer. Around 1905, his wife died after giving birth, and he had already gotten on the wrong side of Paul Kruger's regime, so he went into exile in the veldt. He soon fell in with the termites, digging into their mounds, running little experiments, intuiting what he couldn't test, and spinning stories. He also spent time with baboons, and even adopted one he named Piet who rode in his wagon, took the reins, and wore a gourd as a hat. He worked alone—taking morphine to expand his senses and dull them, too. He apparently lacked some of the usual fears: his biography states that he once injected dried mamba venom into a honey badger to see if it was fatal. I don't know how he got either of those animals to submit to the process. He read some Freud, tried to submit a paper to the Smithsonian about male baboons having sex with each other, and hypnotized his neighbors. Years later his son told a biographer, "In any piece of writing by my father you can always see at work the poet, the journalist, the morphine addict."

I read *The Soul of the White Ant* as a cautionary tale about the way stories shape our understanding of nature. Or just a cautionary tale.

But in spite of all that, Marais was a surprisingly accurate observer of termites.

Continuing the story: his termite "statue" of a queen emits a perfume-like "wireless SOS" into the air. Today we'd call that a pheromone—a specialized chemical hormone that signals to males that the queen is fertile. In Marais's day, pheromones had not yet been discovered, yet his description of a perfume is not far off: in 2012 the researcher Kenji Matsuura analyzed the queen pheromone of another type of termite and found it was composed of a chemical found in pineapple and an alcohol that is a component of the smell of black truffles. Eau de Termite. Understandably attracted to this delicious scent, a male may manage to land unsteadily nearby, scratch off his own wings, and then race off with the female across the damp ground. "His haste is terrible and irresistible" as they scuttle off in search of a spot to dig a burrow where the two will mate.

At first the two termites will be alone in their dark hole. The modern researchers Christine Nalepa, Theo Evans, and Michael Lenz have written that termite parents bite off the ends of their antennae, which may make them better at raising their young. Antennae give termites lots of sensory information, and biting off the segments toward the ends could reduce that stimulation, making it easier to live in a tiny burrow with a few million children. (It may be analogous to how human parents, numbed by sleep deprivation, can deal with toddlers' tantrums without losing their minds.) It's also likely that removing some antenna segments alters which genes get expressed, changing the specific mix of chemicals that Ma and Pa Termite produce.

Soon after, the queen gives birth for the first time. Marais:

> For a long while she stands on the place where the eggs are to be deposited, before she begins laying. Her body is in constant movement. The antennae sweep in circles and her jaws move ceaselessly. Occasionally she lifts the hinder part of her body in just the same way as she did when she was sending her first signal to her mate. Two or three times before the eggs actually are laid, she turns round and looks at the ground as if she expects to find something there. With the actual laying of the eggs the bodily contractions increase tremendously. When the first batch is laid, she turns round once more and examines them long and carefully. She touches them gently with her jaws and front legs, and then she lies motionless beside them for a time. What does it all mean?

That line always cracks me up: "What does it all mean?" Is the termite queen the torch singer Peggy Lee, asking if that's all there is to an egg?

Speaking for Nature, Marais tells the queen: "Although you will apparently be an immobile shapeless mass buried in a living grave, you will actually be a sensitive mainspring. You will become the feeling, the thinking, the seeing of a life a thousand times greater and more important than yours could ever have become."

AFTER SHE HAS laid her first eggs, the queen cleans them often to remove harmful fungi until they hatch as nymphs about three weeks later. The nymphs will molt, grow, and develop, but under the influence of the queen's pheromone, most of them won't fully mature, remaining permanent stay-at-home preteens—eyeless, wingless helpers. Males and females alike will spend their time gathering food, tending eggs, building the nest deeper into the ground, and eventually tending a fungus. As the family grows bigger, some morph into soldiers: possibly, they receive special cocktails of chemicals while feeding from their parents' anuses, causing a change in both behavior and the shape of their heads, which grow larger, dark-colored, and hard in a distinctive way, depending on their species. Thereafter they must be fed by their siblings the workers. Soldiers appear to return the favor by dosing the colony with antimicrobial secretions that help it resist disease.

Over time, in the small smooth dirt room where she lives, the queen's body becomes "physogastric," her abdomen swelling to the size of my thumb, constricted by taut black bands remaining from her old exoskeleton so she looks like a soft sausage that's been carelessly bound with string. Or, as Marais puts it: "an unsightly wormlike bag of adiposity." Her head, thorax, and legs remain tiny. Immobilized, except for the ability to wave her legs and bobble her head, she lays eggs at the rate of one every three or so seconds. The king stays by her. Her children lick off the liquid that appears on her skin, feed her, and care for the eggs.

Or at least, that's life for some *Macrotermes* queens. There are, however, at least three thousand named termite species, and thus at least three thousand ways to be termites. Some have multiple queens; some have cloned kings or queens; some are, improbably, founded by two male termites. One species doesn't really have workers. Different species eat wood, others eat grass, and some eat dirt. *Macrotermes* tend a fungus, but most others do not. All termites, though, live in their own version of a big commune.

Marais called the termite mound a "composite animal," uniting the millions of sterile workers, the soldiers, the fat queen, and the king with the dirt structure of the mound itself into a single body. "You will need to learn a new alphabet," he warned his readers before leading them in. The hard-packed dirt on the outside of the mound, he said, is a "skin" constructed by termites, which build passageways inside that allow the mound to breathe—like a lung. The organism's stomach is the symbiotic fungus that sits in catacombs under the mound, digesting grasses delivered by termites. The mound's "mouth" can be found in the hundreds of foraging tunnels the termites construct through the surrounding landscape. Because they carry nutrients and rebuild the mound, the sterile workers resemble blood cells. The mound's "immune system" is the soldiers who rush to defend the space whenever it is invaded. To Marais, the queen was no Victoria, but instead a captive ovary, walled into a chamber no bigger than her swollen, sweating body. Marais imagined that eventually the mound would evolve into a being that could move across the veldt—very slowly in its dirt skin—a monster hybrid of soil and soul.

Marais's insight wasn't original to him, and many scientists had taken to calling such social arrangements of termites, bees, and ants "superorganisms." The originator of the term was the entomologist William Wheeler, the founder of the study of ants in the United States. In 1911 Wheeler wrote an article called "The Ant-Colony as an Organism."

For a time, superorganisms were all the rage. The concept dealt neatly with what Darwin had called the "problem" with social insects. Darwin's theory of evolution proposed that natural selection worked on individuals and the fittest individuals bred with others similarly fit to their ecological niche, while the less fit were less likely to reproduce. The problem with social insects was that while single termites seem to be individuals, they do not function as such. Only the queen and king of a colony breed, so who was the "individual"? By declaring the whole colony the individual, Wheeler said its members made up "a living whole bent on preserving its moving equilibrium and its integrity."

In the late 1920s and early '30s, the paradigm of the superorganism grew colossal, verging on physogastric. Instead of studying individual trees, biologists studied forests as superorganisms. By 1931, the concept snuck into popular culture when Aldous Huxley reportedly based the dictatorship in *Brave New World* on humans as social insects, with five castes. Wheeler proposed that "trophallaxis"—a word he invented for the way insects regurgitate and share food among themselves—was the secret sauce, the superglue of societies both insect and human, and the foundation of economics.*

Even during the superorganism's heyday, Marais was alone in his assertion that the mound had a soul. His description of it was fantastic: "The functioning of the community or group psyche of the termitary is just as wonderful and mysterious to a human being, with a very different kind of psyche, as telepathy or other functions of the human mind which border on the supernatural," he wrote. Sometimes Marais seems to be working some new ground between Freud and Darwin, while at others he seems to be trying to suss out the obsessive behavior of his own brain. But in that "soul," he was subversively tweaking Descartes's famous definition of animals as "soulless automata," and challenging his readers to ask how different we are from animals, or even insects.

If a child in a fairy tale went into a termite mound looking to save his own soul, you might expect him to come to ruin. And so it was for

---

*Wheeler supported Herbert Hoover's presidential campaign because Hoover had been in charge of shipping American food to Europe after the end of World War I. Wheeler considered this a giant intercontinental act of trophallaxis, according to the historian Charlotte Sleigh.

Marais. He believed that the newspaper articles that formed the basis of his book were plagiarized by the Belgian Nobel Prize winner Maurice Maeterlinck, who published a book called *The Life of the White Ant* in 1926. From South Africa Marais fought for recognition but eventually gave up. He struggled with addiction and depression. In 1936 he borrowed a friend's shotgun "to shoot a snake," and turned it on himself. He didn't die with the first shot and had to fire a second.

Marais's observations had not brought him any expectation of benevolence: "If Nature possesses a universal psyche, it is one far above the common and most impelling feelings of the human psyche. She certainly has never wept in sympathy, nor stretched a hand protectively over even the most beautiful or innocent of her creatures. . . . Pitiless cruelty, torment, and destruction of the weak and innocent. The thief, the assassin, the blood-stained robber, these are her favorites."

WORLD WAR II turned out to be great for invasive termite species, but terrible for the concept of the superorganism. With the shipment of goods and munitions around the world after the war, the Formosan subterranean termite was transplanted from Asia to Louisiana and other southern U.S. states and began to spread in massive supercolonies.

In the aftermath of the atom bomb and with the discovery of DNA's double helix, biology turned away from grand holistic theories toward discrete parts like genes and molecules. This shift reordered biology's priorities and perspectives from the forests back to the trees. In the backlash, the superorganism became "a pariah," according to the complexity theorist Peter Corning. By 1967, the scientist E. O. Wilson said the inspirational concept of the superorganism was nothing more than "a very appealing mirage." And forget the soul: "The organism is only DNA's way of making more DNA." Later he described one of his magnificent leaf-cutter ant colonies as "an organic machine." For decades, scientists studying termites and social insects avoided using the word "superorganism" in their papers.

But by 2009, when I first went to Namibia, it seemed that the concept had merely been dormant, because it was suddenly a topic of discussion again. I couldn't tell whether superorganisms were an idea with actual scientific merit or a rich metaphor that raised important questions about social insects. In any case, Wilson himself had recently written a book titled *The Superorganism* with his collaborator Bert Hölldobler.

Although I was driving to meet Scott Turner to write a story about his own research, I also hoped, on a personal level, to find out what he thought about the idea of superorganisms and the work of Eugène Marais.

Marais's wild tale, for all its flaws, still provokes conversations. One day in 1999, for example, the evolutionary biologist Richard Dawkins and the evolutionary psychologist Steven Pinker met in London to debate the question of whether science was killing the soul. Unsurprisingly, the two agreed on too many points to have much of a debate. But then Dawkins brought up *The Soul of the White Ant* while talking about the design of the human brain. He explained that the book describes termites as individuals working for the single unit of the mound, following rules, so that the mound has emergent properties—it's more than the sum of its parts. And then he wondered if the human brain is likewise made up of many termite-like "mindlets," all cooperating and creating the illusion of one unitary mind. He ended on a plaintive note, as though turning his own brain inside out: "Am I right to think that the feeling that I have that I'm a single entity, who makes decisions, and loves and hates and has political views and things, that this is a kind of illusion that has come about because Darwinian selection found it expedient to create that illusion of unitariness rather than let us be a kind of society of mind?"

After so many hours of flying and driving, my mind had given up all pretense of unitariness. I struggled to focus on the straight highway while ignoring the termite mounds enthusiastically motioning me onward.

THREE

AN INCONVENIENT INSECT

LATE THAT AFTERNOON I reached Otjiwarango, a dusty grid of a town with two big grocery stores and a street named Hindenburg. Scott was waiting for me at a gas station. Bearded, probably around sixty, he bent over to look at me with the posture of a question mark. He navigated as I drove out of town to Omatjenne, the forty-two-thousand-acre government research farm where he had been working for nearly a decade. I slalomed the Toyota around the mud pits on the farm's long dirt road while Scott offered coaching and spoke affectionately about warthogs and kudu and guinea fowl. He was intensely pragmatic, no mention of Mars. And while his manner was a formal version of jolly, the affable tour guide, I knew I was being closely observed.

Over the next ten days I came to appreciate Scott's capacity for cerebral observation. Ask him a question and he'd blink behind his gray

gradient lenses, offer a long, nasal *hmmmm*, and begin a Socratic interrogation of the topic in a guarded tone. I never heard him answer a
question without giving it thought. Often, like a chess player, he'd already thought three steps beyond. Sometimes, as he stood looking at
a mound, or a termite, it seemed like *thinking* could be a physical thing:
something that could be heard or measured. He never spoke of his
emotions but they were easy to read in his shoulders, which often
hunched with defeat and occasionally folded even further downward
into abjection.

The lab and living space were in a neat single-story farmhouse
made of cement blocks. The grass around the farmhouse was about
four feet tall, and termite mounds stuck up above it like a pinkish
archipelago in every direction.

Scott's lab was in the farmhouse's largest room. It was furnished
with derelict camp furniture and a hodgepodge of equipment: a blowtorch, petri dishes, cylinders of oxygen and carbon dioxide, a mass flow
controller, a periscope made from a mirror and PVC pipe, empty margarine containers, an eighty-pound bag of concrete, and a light table
propped up on James Ellroy's *The Black Dahlia*. The walls were decorated with spreadsheets with headings like "60 termites, 10 July 2006."
A pile of laptops, cords, and adapters sat safely at the end of the lab,
away from the mud that was often tracked in from outside. You could
do anything you wanted here as long as you were capable of building it with a soldering iron and an old Sunshine D margarine tub.

We got into Scott's battered VW van and took a ten-minute ride
through the fields, passing hundreds, maybe thousands of mounds.
Some of them showed signs of Scott's experiments over the years:
they'd been drilled, cut in half, invaded with a homemade scope, or
sprayed with water, only to be rebuilt later in some altered way by the
remaining termites. The broken mounds were a memory of all the
questions that he had asked, trying to figure out how termites built
them and why, and also a testament to the termites' resilience. The
grasslands appeared to be nearly empty, except for a few cows, but
dozens of workers lived on the farm.

We stopped at an open shed with something white glowing lightly

under the roof. It felt like a shrine. Here Scott had once filled a termite mound with plaster of paris and then washed away the mud. What was left was an eerie sculpture of curving white pillars of varying thicknesses, reaching about twelve feet skyward to end at the same rakish angle I had observed on the mounds while driving up. This new thing was the exact inverse of a termite mound—present where the termites had removed the dirt, absent where they had built it up, mouthful by little mouthful. It was a beautiful object, with large sinuous plaster shapes revealing the smooth tunnels the termites had excavated in the middle of the mound surrounded by a lace of smaller ones.

The sun was setting and I was very tired; greenish twinkles skittered across the white surfaces of the not-mound and it seemed to animate slightly, the plaster forms echoing bleached bones and blood vessels with a vague but undeniable personality.

The mound, Scott explained, looks like a fixed structure, but it is really a dynamic process. It's always falling down and rising up at the same time—much the way our bones are continually disassembled and repaired as we stand on them. And as with our bones, the mound's structure is organically coupled with its functions. Rains and animals knock down parts of the mound and the termites rebuild, with a preference for the warmer sunny side. Scott spent years taking time-lapse photos of mounds every day at the same time. Run together like an animation, the photos reveal protean mounds, wiggling and waggling that finger of dirt toward the sky. In the four or five years it took for termites to form a colony and build this mound, that process of rain-driven falling and sun-driven rising came to approximate the average zenith angle of the sun at this latitude: 19 degrees from north.

The mound, then, became something entirely beyond Marais's fairy tale: it was a computation of its own position in the solar system, relative to Earth's latitude, in mud, with termites as the calculating agents. So there was the answer to the first riddle of the north-pointing mounds.

Termites are an inconvenient insect, Scott said. Over the past thirty years, scientists have sketched out a grand mathematical theory of how

social insects have evolved. While it works for ants and bees, it does not work for termites. More generally, he has watched the cryptic logic of the mounds ruin one beautiful intellectual theory after another, leading him to a set of unwieldy and unconventional ideas. But for Scott, the *Macrotermes* present a more emotional inconvenience, because they take him away from his family in Syracuse, where he's a professor.

Scott was trained in physiology, a strain of biology that has a long history—back to the grand holistic thinkers of the 1800s. When biology turned reductionist in the 1950s, physiologists refocused on finding biology's little machines. Physiologists can tell you about the design of the giraffe's heart that allows it to lower its head to the ground without fainting. They will tell you that tuna have heaters in their brains, giving them the slightest advantage in outwitting predators.

Scott dropped out of college to work in construction and returned to school with a large set of skills for fabricating structures and devices, an interest in biology and physics, and a constitutional irritation—like a grain of sand in his viscera. He has a basic impatience with questions that are too small and safe. Early on he put alligators in wind tunnels to see how their body shape helped them regulate their temperature. "I had a collection of fifty alligators," he said. "But you know . . . putting an alligator in a wind tunnel, for crying out loud! Who does that?" In 1985 he got a postdoc position at the University of Cape Town, South Africa. He had gone there in pursuit of alligators, but started studying the physics of egg-warming—all very mathematical and reductionist. Apartheid South Africa was under a scientific boycott at the time, so he started building his own electronics. He also fell in love with Deb, a vivacious South African, got married, and had a child on the way.

In the late 1980s, he found a job teaching the principles of physiology at a nursing school in a remote town near the Botswana border. He thought that the three-foot-high termite mounds around the school would be a good way to demonstrate a well-known theory about how open-topped mounds function. The going hypothesis was that wind blew across the top of the hole, inducing the hot air deep in the mound to flow out the top, while cool air rushed in from the sides. He found a broken gas leak sensor, repaired it, and bought a canister of propane to demonstrate that the mound worked in an organized fashion. When he put the propane in the mound, however, it behaved unpredictably—sloshing around sometimes, rising others, seemingly dependent on whether the winds were steady or gusty. He became interested in termite mounds, and their mystery deepened when he moved his work to Namibia. The closed-top mounds found in Northern Namibia had been famously described as working like chimneys—with buoyant warm air in the nest rising through the mound and out the top—by a physiologist named Martin Lüscher in the late 1950s. When Scott measured flow in the closed mounds with propane, it sloshed again.

And so there went a beautiful theory. But what would replace it? It would take Scott years of experiments to show that mounds could work a bit like lungs, with interconnected chambers taking advantage of fluctuations in wind speed. Air moves back and forth through the porous dirt skin of the mound by two systems: in big puffs driven by buoyant gases rising from the hot fungus nest (like the sharp intake of breath from the diaphragm), and in small puffs, the way air wheezily diffuses between alveoli in your lungs. Scott suspected that the termites themselves circulated air as they moved, like mobile alveoli. This insight didn't just prove Lüscher wrong; it was an entirely new way of thinking about the problem. The mound was not a simple structure where air happened to move, but a continuously morphing complex contraption consisting of dirt and termites together manipulating airflow. Somehow, while distracting himself with a broken gas sensor, he'd found the biggest animating question of his career.

Early in his work Scott had done experiments with propane and later put small plastic beads in the mounds to see how termites moved particles. More recently, he'd been trying to get a deeper, more conceptual sense of what they were doing by working with computer game designers and an engineering professor named Rupert Soar.

I met Rupert when we went back to the farmhouse for oryx steaks and small squash roasted over a campfire. He wore a mud-spattered argyle vest and longish shorts, and stood near the fire drinking beer along with a boisterous group of similarly dirty British men. There were nearly a dozen people staying in the farmhouse, including Scott and some grad students, and dinner had the feeling of a salon. Though he looked exhausted, Rupert had the ability to pull whole thoughts, paragraphs, out of the muck. "The concept of construction is very human and it ends with occupation. The process with termites is ongoing so that the structure morphs over time. It's never finished."

Termites who spend a year building an average mound ten feet tall have just built, in comparison to their size, the Empire State Building. Those who build taller mounds, at nearly seventeen feet, have just built the Burj Khalifa in Dubai—2,722 feet and 163 floors of vertigo—with no architect and no structural engineer. Such unthinking, seat-of-the-pants design is not possible for humans, who required squads of professionals, advanced equipment, and seventy-five hundred people working for six years to build the Burj Khalifa. Rupert hoped to harness the powerful constructive groupthink that comes from the tiny mouths of termites and their even tinier brains to build structures in remote environments like Mars. But there were issues: termites, he said, engineer to the point of collapse—their mounds only need to stay whole 51 percent of the time to survive.

Rupert had recently left his job at a university to devote himself to exploring the possibilities of termite-inspired buildings. He'd brought the crew here to make a plaster cast of a mound in the field in the hope of selling such casts to museums. This, I later realized, is exactly the sort of vexed inspiration you'll get if you stare into termite mounds long enough. If the team could pull the plaster mound out of the dirt, they'd need to load it into a truck and drive it to Windhoek without breaking it.

I mentioned that the road—the B1 that I'd just driven up—seemed plenty straight and flat for the task. Someone explained that it stretched from South Africa nine hundred miles north to Angola, and that it had probably been used to transport military supplies for South Africa's Angolan wars between the 1960s and the 1980s.

I found a mattress in the back of the house and went to bed having learned the answers to both of the afternoon's riddles: the north-pointing mounds and the very straight road. What I didn't know that night was that the answers would become puzzles themselves, leading me further into termite territory.

FOUR

INTO THE MOUND

ONE MORNING, WHEN SCOTT STOOD in the kitchen mixing fluorescein dye in a large Coke bottle so that it glowed an ominous yellowish green, there was a loud mechanical shriek in the yard.

A tall white man got out of a dust-covered Land Rover. Whatever was holding the Land Rover together was also at work on his clothing: one sandal hung by a broken strap, giving him a slight limp; both his shorts and his shirt were ripped. His hair was brown and gray and it stuck out at angles from his head in a kind of fan. His face was deeply sunburned and he squinted as though the light was too bright, which I first assumed was a habit or a tic but came to believe was a way to keep his glasses from sliding down his nose. He was an entomologist, the chief curator of Namibia's National Museum, and the reason Scott was doing work on the farm.

And his name was Eugene Marais. He said he was no relation to the other, but like him he had stories to tell about lions and snakes. Later in the day he handed Scott a scorpion. In his youth, Eugene was sent to Europe by the Namibian government with a large case full of entomological specimens to take to the big museums in the hopes of getting definitive identifications. He had to wear a suit for the task. It was easy to imagine the big case of specimens; much harder to imagine the suit.

The three of us got into the van with the Coke bottle, which now glowed like a kooky prop from a Disney movie about professors. Under a black light fluorescein becomes bright green, and Scott wanted to use it to trace where termites carried water. On the way to find a mound to dump it into, we drove past the twelve mounds that the grad students photographed daily. Whenever dirt fell from them in the rain, a student scraped it up, tagged it, dried it in a nearby cowshed and weighed it. In this manner Scott had gotten a basic idea that 11 pounds of termites could move about 364 pounds of dirt in a year.

Each observation served to embellish a story, a big pile of hypotheses. Termites suck water into their own bodies, sometimes taking up a quarter or even half of their body weight in water. They also grab soupy mud balls and move them to drier parts of the mound. For every pound of dirt the termites moved, they also carried nine pounds of water, meaning that in a year those termites were also moving thirty-three hundred pounds of water. If you could see the mound on the right timescale, it would be a bucket brigade of water—a very diffuse water fountain, spilling over into the dry land.

BUT WHAT, EXACTLY, were the termites doing? When Scott first started poking around the mounds, he'd assumed that the structures had evolved for the purpose of exchanging gases, much the way gills in fish evolved to be lungs on land. But the more he understood about the volume of water flowing through the mound, the more he wondered if it had evolved as a way to balance water and dryness and maintain the right moisture in the nest, making the gas exchange just a by-product of the flow of water. It was a kind of circular reasoning. He mused, "Does the spire keep the termite mound dry, does it give them a place to move the water to?"

Eugene interjected that this was part of a bigger discussion that he and Scott had been having for twenty years. How do you suss out what it is—dirt, water, or air—that motivates termites?

When biologists try to explain why a behavior evolved, it's easy to veer into "just so stories" that find some neat evolutionary narrative to explain the way things are, much the way Kipling explained how the camel got his hump. But some behaviors are simply by-products of something else. Scott and Eugene discussed whether the flow of water via termite could just be a side effect of some other project the termites were engaged in.

The air had become heavy and clouds were lining up above us in a formation that looked like a row of cup handles. Eugene spotted a baby cobra with tiny fangs swiveling in the grass and shooed it with his broken sandal, chortling another part of his ongoing discussion with Scott: "Nature red in tooth and claw!" This got a guffaw from Scott, who was walking among the mounds with a large auger. He said that the idea of competition as the driver of evolution was misleading. Without being (as Scott would say) "kumbaya" about it, the ability to cooperate is a powerful strategy for survival.

One thing that came up often when Scott and Eugene talked termite was their shared skepticism about the dominant explanation for the evolution of social insects, called inclusive fitness theory. The theory was another attempt to address Darwin's "problem" with social insects—why did they cooperate with each other when competition seemed the surer route to increased rates of reproduction and survival? Perhaps it was the Cold War's focus on the positive aspects of capitalism, but after World War II science came to view the "altruism" of social insects as irrational, evolutionarily speaking. The inclusive fitness theory, though, demonstrated how altruistic bugs could benefit.

The theory was partly inspired by something the British biologist J.B.S. Haldane said in a bar in the 1950s. Haldane stated that he would risk drowning to save two brothers or eight cousins (implying that he would not jump in to save one brother or seven cousins). The gist, for biologists, is that altruism pays off if it saves a majority of one's genes, and thus genes for traits like altruism could spread through a population. In the early 1960s, the evolutionary biologist William D. Hamilton used Haldane's insights as the basis of an equation showing that altruism is a successful reproduction strategy if the genetic relationship between individuals and the benefit they derive from the altruism is greater than the cost of helping others.

This equation seemed to explain the sociality of bees and ants, where females may be super sisters, genetically. The reason for this is that bees, ants, and wasps are "haplodiploid," which means that males develop from unfertilized eggs, with just one set of chromosomes from the queen. Meanwhile, bee and ant females have two sets of chromosomes, and if they have the same father, they can share three-quarters of their genes with their sisters. So for ants and bees, the life of all-for-one altruism made sense as a genetic strategy. In 1965 the entomologist E. O. Wilson was on a long train ride when he experienced what he called a "paradigm shift" in his thinking about the importance of this mathematical explanation. Other scientists agreed, and for several decades inclusive fitness was the dominant theory, inspiring research and many papers.

But the theory was less convincing for that "inconvenient insect," the termite. Unlike bees and ants, termites are diploid, and brothers and sisters generally develop from fertilized eggs, so the equation did not fit the way it did for the other social insects. By the time I was in Namibia, Wilson—who had by then returned to the concept of the superorganism—was having second thoughts on inclusive fitness theory, too. In 2010, with the biological mathematicians Martin Nowak and Corina Tarnita, he published a paper revisiting and refuting the idea of that mathematical genetic explanation for altruism, in favor of a concept called "group selection" that argued that cooperating kin had more surviving children. To riff on Haldane: Altruists simply have lots more brothers and cousins that survive. In response, more than one hundred scientists wrote a letter refuting the paper.

Scott climbed on the base of a tallish mound, braced his feet, and began turning the auger into its gray cement-like skin. From ground level he looked both prophetic and comic. "We'll know we got there when we smell the fungus combs," he said. The scent of bread and dirty socks soon wafted out of the hole. Scott inserted a piece of pipe. The fact that it was dry in the mound when all the fields were wet suggested to Scott that the termites were frantically shuttling moisture from one place to another. "They're moving huge amounts of soil, and right now they're behind. They won't catch up until after the rains stop. They don't sleep; they work all the time." He thought maybe they were dragging the wet soil to the top of the mound, where it was more likely to get dried out by the wind. And that gave him another idea about a different experiment. But he still needed to do this one: he dumped the contents of the glowing soda bottle into the dark hole in the mound. If using the dye as a tracer worked on the first pass, he'd devise a more tightly controlled experiment and repeat it several times. Few scientists have the freedom to work this way now; usually they're constrained by grants and by limits in time, labor, money, and the need to publish. I wrote in my notes: "This is all very ad hoc."

Later we passed Rupert and his men working on the plaster cast of their mound. They had poured five tons of plaster of paris into the top of a mound, embedded a chain in the plaster, hung the chain from a gantry, and cleared away the dirt at the base of the mound. What I could see was about the size of a grand piano, but looked like an enormous turnip above a reflecting pool of muddy water. Rupert's cheeks were pink with emotion and he smiled grimly as he tried to hose the dirt off the thing. Big ghostly knuckles of plaster appeared under the spray from his hose and disappeared in murky streams. "We're going to lose it. There's nothing we can do," he said. A stoic, he called the whole melting, mocking mess a "Zen experience." He still hoped to salvage the bulk of the mound. The helpful young men crowded toward the hole, and it seemed just a matter of time before the thing fell on someone.

Later in the week we heard a loud rumble, a wet slump, and a splash. Rupert's mound had fallen into its pit. A short piece of plaster and mud, marbled white like a well-fatted ham and about the same size and shape, swung wildly on the chain while the rest of the plaster and mud disassembled into nonsense in the pool below.

EARLY THE NEXT morning a backhoe arrived and Scott directed it to the mound where he'd left the fluorescein dye. The backhoe's great blade came down on the top of the mound with a hollow whomp, the first note of a funny little concert. Half the mound fell away with a tumbling clinking clatter—as the shards hit different layers of cured mud they played a tune like a soft xylophone. We pushed in close, enveloped by the familiar smell of socks and bread.

What was left of the mound was a ruined hierarchy. Dirt shards and fungus combs and sculpted mud plinked downward, while termites ran every which way, at first as a sort of gauzy net. Soon they had organized themselves into small streams, and within ten minutes those streams had consolidated into rivers of running bugs. As order was restored, I could see the elaborate scheme of tunnels, rooms, chambers, and fungus hidden under the dirt exterior. The spectacle was genuinely awesome—as in jaw-dropping and appalling. It's easy to understand why the ancient San people made thousands of rock drawings said to be of termite mound interiors that may have been seen as gateways to the spirit world.

The top of the mound was hollow, with the wide vertical tunnels that I'd seen cast in plaster the first day. The interiors of these tunnels were very smooth, made by fine craftsbugs, and they segued in and out of each other in ropy vertiginous columns like a sloppy braid. Termites make the mounds by first piling up dirt and then removing it strategically in the tunnels, creating shapes that feel "right" to them. Eyeless, they use their antennae to feel for smoothness, and in the big tunnels they remove everything that is rough. They may even *hear* the tunnel's shape. Rupert once put sophisticated microphones in a mound and found that the big vertical tunnels have a resonant frequency, like an organ pipe.

Termites are often compared to architects for the way they build their mounds, but that is misleading because they don't have plans or a global vision. What they really have is an aesthetic, an innate sense of how things should feel. The termites are, in a sense, millions of Martha Stewarts: constantly remodeling.

When the top of the spire was first ripped off, there were just a few termites in the solitary tunnels at the top, probably listening to the clopping of their own six feet. But cutting into the top allowed in lots of fresh air at once, and activated an alarm system. Some termites ran away from the hole, agitating their brothers and sisters so they could help with repairs. Thousands of worker termites followed the smell of fresh air to find the hole, carrying balls of dirt in their mouths. Within minutes of the backhoe strike, streams of termites canvassed the broken side of the mound, moving in a frantic start-stop pattern like a shaky old animated cartoon.

I leaned in further and could see that each termite put its ball of dirt down on a ball left by the previous termite, wiggled his or her head, perhaps to get the ball to stick, and then backed away. Where there were two balls there were soon twenty and then two hundred, then two thousand. Some of these stacks joined up with other stacks at the perimeter of the breaks in the mound to form little bumpy, frilly walls. Once the area was walled off, Scott said, the signal from the fresh air would stop and the termites would fill the internal space with more dirt balls and small tunnels, making a sort of spongy layer. Later they'd either block it off entirely or they'd hollow it out and remodel it. Nothing ever stays the same in termite town.

"Bastard!" Eugene yelled. The fresh air had also drawn the dark-headed soldier termites to the edges of the break, where they launched themselves blindly outward while snapping their mandibles. *Macrotermes* have two sizes of soldiers. Major soldiers are much bigger than the workers, and they have square maroon heads sporting long, sharp, curving blades that scissor together with great force, even as they leap to their deaths by the hundreds. One drew blood from Eugene by cutting through his leather gloves—altruism in action.

The backhoe came back in for another swipe, taking away the dirt below the mound to reveal the system of horizontal galleries, tunnels, and chambers where the termites live. It reminded me of those diagrams of cruise ships, visualized from the side, with small rooms packed together in a strict hierarchy of function and status from ballrooms and cafeterias to VIP staterooms and steerage bunks. The colony's hierarchy is not money, of course, but the things that enable its survival: reproduction, child care, food supply, and food processing. With walls and floors the texture of tortilla chips, some rooms are large, with vaulted ceilings. When I looked closely, I could see that they were not so much rooms as places where many foraging tunnels crossed, like the grand concourses of old train stations. Deep within this area was a small capsule where the king and queen lived, making eggs, which were carried to nearby nurseries.

As I looked into the mound and saw their smooth art-deco walls, I scrawled in my notes excitedly, "Termites live in the future! The Jetsons." But later I realized there was something wrong with that. Sure, the mound had the futurist architecture of the old TWA terminal at JFK—or *The Jetsons*—but what I was really looking at was something well beyond the future of flying cars.

BELOW THE MOUND lives the fungus, digesting grass. All termites use symbiotic collectives of bacteria and other microbes to digest cellulose for them, but *Macrotermes* outsource the major work to a fungus. In some senses the fungus functions as a stomach, but it also has power reminiscent of the Wizard of Oz.

Under the mound and around the nest sit hundreds of little rooms, each containing fungus comb. This comb is made of millions of mouthfuls of chewed dry grass, excreted as pseudofeces and carefully assembled into a maze. This comb roughly resembles graham cracker pie crust, though it varies in color from delicious beige to decrepit black. The termites inoculate the comb with a fungus that they have been cohabiting with for more than 30 million years.

You can pull the fungus combs out of their little rooms as if you were pulling drawers from a doll's wardrobe. The comb maze wiggles like the folds of a brain, with the hard, wrinkly piles of chewed grass making the gyri and leaving sulci-ish gaps in between. This is not an accident: as with a brain, the comb design increases the surface area of the structure. Within the gaps are what look like tiny white balloons, which is the fungus blooming. There is nothing accidental about *this* relationship either, or the construction that holds it: the details are so fine we can barely take them in. The bottom of the fungus comb stands on peg-like legs, little nubbins that hold it up just enough to let air circulate through. One of the grad students beat a small stick against the floors of the fungus galleries, playing something that was almost a tune.

The symbiotic relationship between *Macrotermes* and the fungus is tight: workers scour the landscape for dry grass, quickly run it through their guts, then place and inoculate each ball to suit the fungus's picky temperament, tend the comb, and snarfle the fungus and its sugars before distributing the goodies to the rest of the family. Then the workers run off to gather more grass for the fungus. Termite and *Termitomyces* fungus are so interrelated that it's hard to tell where the mushroom ends and the termite picks up, but within their codependence is some sort of frenemy-type rivalry.*

Prejudiced by our human sense of a hierarchy of the animate termites over inanimate mushrooms, we'd be inclined to believe that the termites control the fungus. But the fungus is physically much larger than the termites in size and energy production: Scott estimates that its metabolism is about eight times bigger than that of the termites in the mound. "I like to tell people that this is not a termite-built structure; it's a fungus-built structure," he says, chuckling. It is possible that the fungus has kidnapped the termites. It's even possible that

---

*Fungi are capable of deliberately tricking termites. One invasive fungus in termite colonies in the United States and Japan pretends to be a termite egg, going so far as to secrete the chemical lysozyme, which the termites use to recognize their eggs. For reasons that are not clear, colonies filled with impostor "eggs" are no less healthy than those without them.

the fungus has put out a template of chemical smells that stimulates the termites to build the mound itself. As I peered at the white nodules, I began to sneeze violently, sometimes with big gasping whoops, and something—it's hard to even call it a thought, but a particle of one— flitted through my subconscious before flying out of my nose: The fungus is very powerful.

My admiration for the fungus only grew when I learned that Namibian farmers estimate that every *Macrotermes* mound—which contains just eleven pounds of termites—eats as much dead grass as a nine-hundred-pound cow.

SCOTT POINTED OUT a rind of crystalline material in the dirt below the mound. He said it might be calcrete, a form of calcium carbonate capable of holding water the way a saucer under a teacup holds spilled tea. Calcrete forms when water, calcium, and methane combine. Calcium is often found in the soil, and termites make methane in their guts when they eat grass, so it wasn't inconceivable that termites had created this dish under their mound.

I had followed along with the logic until this point—and I had stayed away from the "just so stories"—but now I was stumped for a follow-up question. It's one thing to imagine how termites might randomly place dirt balls to build lung-like mounds that offer ventilation, or how termites' response to a surplus of water might lead to a behavior that results in a mound. But a water-conserving dish produced by a chemical reaction? This was a more sophisticated series of events than I was ready to attribute to the termites. It was hard to even phrase the question. Would such a saucer-type thing be deliberate? How would a dish emerge from random interactions and then be selected for by evolution? Would termites "learn" to make the dish, or would it be somehow encoded in their genes?

Scott replied, "This is the big dilemma in evolutionary biology. It's kind of a faux pas to speak of deliberate or intentional. The proper way to think of it is as a series of genetic experiments that either allow the termites to survive or die." That made sense to me. He continued, "The issue is also that humans are intentional creatures and we may be imagining that termite colonies are deliberate simply because we're human. But I think that if you deny the intentional behavior, you're missing biology. The challenge is to come up with a credible theory of intentionality." He paused to deliver the logical endpoint of his argument. "And you also have to wonder what it is that allowed us to become intentional beings." Looking at termites, I started tripping over the gyri and sulci of my own brain, stuck in a self-referential maze.

Late in the day Eugene used a pickax to pop the royal chamber out of the nest—the whole complex was the size and shape of a squashed soccer ball, but made of hard-packed finely grained dirt. He cracked it open, revealing the king and queen in a hollow space the size of a cough drop tin. The chamber had holes on the sides, allowing air and smaller termites to pass through. The king was large and dark compared to the workers, but the queen was huge—as big as my finger. Her legs and upper body waggled but barely budged the fluid-filled sac of her lower body, which pulsed erratically, as though she was a toothpaste tube squeezed by an unseen hand. Her skin was shiny and translucent and the fats inside her swirled like pearly cream dribbled into coffee.

Everyone shuddered: the queen is viscerally repulsive. She offends our sensibilities and she is monstrous. I think the first stimulus to shudder is a reflexive reaction to her body's pulses and swirls. But then a more intellectual sense of her horror kicks in. "She's not a queen; she's a slave," said Eugene. Captive of her body, of her children, of the structure of the mound she conspired to build.

Even then, the queen's more shocking aspects are hidden from us. Her truly stupendous fertility—creating millions of eggs over as long as twenty years—is something we can only infer. Some species of termite queens can clone themselves by producing eggs with no entryways for sperm, which then mature into sexual queens with only their mother's chromosomes, duplicated inside the egg nucleus, to furnish a full set. Imperfect copies of the queen, these knockoffs are good enough to get the job done. Parthenogenesis allows the queen to live, in bug years, pretty close to forever.

And yet we *do* refer to her as a queen. I wondered why. Eugene said that when early European naturalists looked into beehives and termite mounds, they saw the monarchies they came from—with workers, soldiers, and kings and queens. It was misleading, he said, and kept us from really understanding what was going on with termites at all.

When I got home, I looked this up. Eugene was correct. Peering into beehives in the 1500s, naturalists literally saw Europe and its political structures in miniature. For two hundred years, they generally didn't describe queen bees as "queens"—that is, females—because they believed only a male king could be head of such a magnificent insect state. It wasn't until the 1670s that Jan Swammerdam, a Dutch anatomist with a microscope and little interest in human monarchies, pinpointed the queen's ovaries and got everyone to agree that she was female.

This was part of a long-standing tradition of seeing ants, bees, and termites as something like tiny humans in insect suits, living in an exemplary "natural" state. Consider the report that Henry Smeathman gave to the Royal Society in 1781 about the glories of termite civilization: "Their contrivance and execution scarce fall short of human ingenuity and prudence; but when we come to consider the wonderful economy of these insects, with the good order of their subterraneous cities, they will appear foremost on the list of the wonders of the creation, as most closely imitating mankind in provident industry and regular government." Smeathman described the mound as England in miniature, with "laborers," "soldiers," and "the nobility or gentry." He noted that bug nobility were worthless: they couldn't feed themselves, work, or fight, but had to be supported by the others. He saw this as a justification for aristocracy—in insects as in humans—"and nature has so ordered it."

In older texts about social insects, you can see them offered up equally as justifications for racism and as inspirations for utopias.

A hundred years ago scientists said that lighter-colored ants kept darker ants as "slaves," explicitly suggesting that racism and slavery had some "natural" basis. Since the 1920s, scientists have complained that the term "slave" distracts from the actual relationships in complex ant societies, but it is still in use. On the other hand, the Russian zoologist Pyotr Alexeyevich Kropotkin emphatically described social insects as a model socialist state in his 1902 book *Mutual Aid*, and the American feminist Charlotte Perkins Gilman examined the beehive and concluded that females had evolved to dominate the Earth.*

I became fascinated by how often social insects have been presented as "just so stories" for a certain kind of human civilization. I dug up a copy of the 1919 comic speech that William Wheeler, the scientist who originated the term "superorganism," gave from the perspective of a termite king named Wee-Wee. King Wee-Wee told a grand yarn about termite mounds as utopias that practiced eugenics, euthanasia, and benevolent fascism. Termites decided to live rationally as a superorganism without going as far as the ants, who had "a militant suffragette type of society." When a termite stopped being productive, "he" was brought before a board of termite biochemists, who assigned him a number that signified his nutritional usefulness. "He was then led forth into the general assembly, dismembered and devoured by his fellows."

The speech wasn't science, but it wasn't comedy either. It read like a sloppy caricature of Darwinian ideas intended to gross out its audience while uniting us with feelings of intellectual superiority. By the time King Wee-Wee mentioned the "physically and mentally perfect race," I felt sick.

But it got worse. He went on an elaborate satirical riff about using gas to kill the old who didn't contribute to the health of the whole,

---

*Gilman saw women's current low status as a "temporary setback," according to the sociologist Diane Rodgers. She dreamed of a global feminist hive: "Government by women . . . would be influenced by motherhood; and that would mean care, nurture, provision, education. . . . We do find it . . . in the overflowing industry, prosperity, peace and loving service of the ant-hill and the bee-hive."

specifying the insecticide hydrocyanic acid gas. Later, with the name Zyklon B, it was one of the gases used to kill millions of people in Nazi death camps. Perhaps King Wee-Wee's combination of ideology and technology once seemed funny, but that time before the Holocaust is not something I can imagine. After reading King Wee-Wee, I gave up hunting for these stories and went back to contemporary science.

For scientists, the great danger of seeing social insects anthropomorphically is that it obscures their true bugginess. In the 1970s and '80s, when the ant scientist Deborah Gordon began studying massive ant colonies in the American Southwest, scientists described the ant colony as "a factory with assembly-line workers, each performing a single task over and over." Gordon felt the factory model clouded what she actually saw in her colonies—a tremendous variation in the tasks that ants were doing. Rather than having intrinsic task assignments, she saw that ants changed their behavior based on clues they got from the environment and one another. Gordon suggested that we should stop thinking of ants as factory workers and instead think of them as "the firing patterns of neurons in the brain," where simple environmental information gives cues that make the individuals work for the whole, without central regulation.

And so, these days, one scientific metaphor for the inscrutable termite is a neuron in a giant crawling brain.

FIVE

COMPLEXITY IS
THE ESSENCE

A **FEW DAYS LATER,** I sat in front of the farmhouse and watched small finch-like birds dipping in and out of the jacaranda trees with a stream of conversational clacking, screeching, and singing. They were tending a large tangle of sticks suspended like an improvised basket in the tree. It was hard to estimate the nest's size—it was an unpredictable shape and partly covered by the leaves—but it might have been as large as a goat or maybe even a cow. Weaverbirds are famous for constructing massive tumor-like nests in trees, and—along with beavers and termites—are some of the most spectacular builders in the natural world. It's thought that what drives beavers to build dams is the sound of running water. And then, like the game of musical chairs, when the sound of water stops, the beavers reportedly stop building. Biologists are still trying to find and understand which cues from the environment turn termites and weaverbirds "on" and "off."

Scott was sitting in the lab, gazing into a baking pan with a few pieces of dirt in it. His glasses had slid down his nose. He was holding a small paintbrush, which he used to move individual termites from the dishpan to a petri dish on the desk. I asked him if he found it funny that he'd ended up studying termites, after starting with alligators. "There's never been any plan here," he said. "I've been led." The more he thought about social insects, the more interesting and strange they became. "There's an awful lot of structure not explained. But complexity is the essence, and if you don't capture it you're not going to have a hope of understanding it."

NOT LONG AFTER he realized that Lüscher's idea that the mounds worked like chimneys was inaccurate, Scott began to doubt one of the other big termite theories: "stigmergy." A French termite researcher named Pierre-Paul Grassé had invented the term in 1959 to describe how the labor of each termite guides that of the next. One termite picks up a ball of dirt, gets its saliva on it, and then drops the ball of dirt. The smell of the saliva—possibly containing a chemical signal called a "cement pheromone"—provides a cue for the next termite to drop its ball of dirt. As more termites drop more balls on the pile, the smell signal gets stronger and stronger, and this cycle of positive feedback induces more termites to drop their dirt balls. Little towers of balls form and some become pillars, and some pillars become walls, coordinated only by the interactions of the separate termites with their environment, not by a master plan. The pheromone smell wears off over time, and so little spires that don't attract many termites don't get new pheromone and don't grow into walls and pillars. (Kind of like tweets that don't catch on on Twitter.)

What makes stigmergy spectacular as a concept is that it doesn't just describe how termites build; it also makes it possible to predict what will be built. Researchers hoped that modeling the process on computers could lead to understanding complicated and mysterious systems of biological organization.

In the early 1990s, the idea of stigmergy moved away from termites and into computer science as programmers created simulations featuring virtual termites, all running the same program. Following a simple ruleset, these computer termites built walls made of x's instead of dirt, without a coordinator or plan. In this theoretical work, sometimes called swarm intelligence or agent-based modeling, termite-inspired programming demonstrated how very simple parts could create a complex whole. These models seemed to offer analogies for other types of problems. For example, how do hundreds of commuters construct traffic snarls? How do thousands of investors influence the stock market? Solve the heady termite problem, it seemed, and we could solve real problems like routing phone calls, fighting wars, and protecting computer networks from hacking.

Swarm intelligence, the name that stuck for this version of termite-think, had unique insights into the complexities of modern life. Eric Bonabeau, an early leader in the field, put it this way in 2003: "We're reaching a stage in technology where you no longer have a choice: your mindset has to change or you'll be screwed. It's no longer possible to use traditional, centralized, hierarchical command and control techniques to deal with systems that have thousands or even millions of dynamically changing, communicating, heterogeneous entities. I think that the type of solution swarm intelligence offers is the only way of moving forward." But the termite solution to the world's problems was tough to find. Computer termites made nice x walls, but they could not model the really complex three-dimensional walls of real termites, never mind the stock market. By 2009, when I talked with Scott in his lab, theoretical termites were still largely theories, and many of the people interested in modeling termites had moved on.

But the more Scott compared the idea of stigmergy with what he actually saw the termites doing, the more confounded he was. He had come to believe that stigmergy was only one of several mechanisms for building. For example, it didn't explain why the termites deconstructed some things they'd built, and why some tunnels were polished and others were rough. As time went on he questioned whether the cement pheromone actually existed. And so stigmergy started to seem like another gorgeous idea, an abstraction that didn't suit what was really going on in the mound.

Scott picked up each termite by pushing the paintbrush on top of it until the insect grabbed the bristles. Then he shook it off into the dish, where it would lie on its back for a minute, wiggling its antennae, before getting up to run around. I sensed that Scott was wary of talking about himself, so I'd taken a seat across the lab to avoid crowding him. As I interviewed him, I began to notice an odd whistling noise. I picked my head up from my notebook and realized he was blowing on each termite before he set it down. Blowing on the termites made them run around. There is something ridiculous about the whole process of moving the tiny insects one by one, blowing on them, and then contrasting the little bugs to the big ideas they generated. Sorting termites is soothing and repetitive, but the impulse to do it comes from irritation, a constant annoyance. The theory doesn't fit. Staring at a termite is reckoning with our brutish ignorance. How complicated can this little thing be?

When he put termites in petri dishes, Scott was trying to measure their behavior. Over the years, he'd noticed that some mounds were slightly taller and pointier, while others were slightly shorter and flatter. They were made by two different species of Macrotermes—natalensis and michaelseni—that are very similar. Since both species were building with the same soil in the same conditions, the different shapes of their mounds were a natural experiment in the genetic differences between the two species. Scott had been ruminating on this for years by examining the two types of termites in their mounds and in dishes and trying to determine how their behaviors varied. Perhaps there was something about the chemical signals they used in building, so that one wore off faster than the other. Or perhaps the builders of the taller mounds got bored more slowly than the others. Every particle of attention added up and glommed into a bigger, higher dirt ball so that the termites with the longer attention spans built higher spires on their mounds. A mound could also be seen as the product of a finely tuned obsession, multiplied by millions of moments of attention.

And here I was watching a man who'd spent twenty years watching these obsessed termites. A genetic lack of boredom leads you to funny places.

I asked him about superorganisms—is the mound a body? Or was that a fanciful concept? He said he thought the link between the organization of bodies and of termite mounds was homeostasis: cells build bones and termites build mounds that create an environment suitable for their survival. He leaned back in his chair and chewed on the brush. "There are tough questions a scientist has to confront. Why do we say the termite is alive but not the mound?"

He continued: "The mound does things the termite can't do. You need to think about the organism extending outside itself through the mound. It's impossible to be alive without changing the environment you live in." This is not a new idea: organisms create pools of dynamic orderliness—homeostasis—from the disorderly universe around them. But in the mound, Scott sees the termites constructing not only a zone for homeostasis, but through their dirt balls, a sort of external memory, and a sense of where they stop and the outside world begins. This is profound and provocative.

For the past fifty years, biology has considered itself the study of genes, but Scott believes this drive toward homeostasis is important and overlooked. "This is a fundamental philosophical question: What is life? Is it genes? Or is it process?" He sided with process, with organisms changing their behavior and their environment. "As a physiologist I see that organisms drag genes into the future—not the other way around," he said. As he talked he seemed to be laying down cards in an unending game of solitaire, part of an ongoing discussion that he has with himself—and maybe with Darwin—all day long.

The implications of his line of thinking have brought him to a funny point: if the termites in a mound "know" where their boundaries are, if they understand where they maintain homeostasis and where they do not, then the combination of termites, fungus, microbes, and mound constitutes some kind of cognitive system. This cognitive system has some sort of desire—to stay whole—and, in some way very different from our own, an outlook on life. And so in a short conversation we arrived at the strange prospect that a termite mound is a kind of filmy consciousness distributed through the soil, an echo of Eugène Marais's soul or psyche.

Scott wanted to build a better conversation between the theory and the experiment, the grand purpose and the small details. Now he had a new plan: understand what the mound is "doing," and use that to understand what cues the termite uses to build, and then try to find some programmers who could replicate these processes in a video game or a robot.

Scott hunched further over the petri dish. "The thing about living systems is that they have these essential qualities of wanting to do something." He used finger quotes around "wanting."

I didn't know where these new termite ideas would lead Scott. And neither did he. Science is often told as a long story of wins and brilliant deductions, leading to the invention of the transistor, or—via evolution—a beautiful triangular beak, perfect for splitting seeds. Failure, though, is at least 49 percent of the story.

Throughout our conversation the weaverbirds continued clacking in their nest, expanding its zone of order stick by stick so it hung in the air above us like a thought balloon.

SIX

BECAUSE THEY ARE SO SWEET

TERMITE ONE'S UNGAINLY OVAL HEAD glowed chartreuse while its spindly legs fell off into violet under the black light. Termite One tilted his or her big head toward Termite Two, who approached darkly from the left. They locked mouthparts. Fluorescent color drained from Termite One, and Termite Two began to glow from within. It was a decadent, Day-Glo, Warholian scene: eusociality as some kind of incestuous insect rave.

Scott sighed, placed his cursor over Termite Two, and clicked his mouse. He'd been mapping the glow since five in the morning, reviewing time-lapse photos taken every thirty seconds. Outside, the dawn was a glorious lavender, but he didn't notice. Click by click Termite One transferred dyed water to other termites in the dish. *Hmph*, Scott sniffed. "More kissing." He clicked and advanced the video, muttering, "One point five minutes later and their mouthparts are still touching." The dye spread across the dish, transferred from mouth to mouth, causing termite after termite to wink on. This wasn't only about water, though. They tilted their heads; they ran their mouthparts over the carrier's antennae, grooming them. They stroked the water bearer's flank. They even seemed to grovel.

As the yellow spread across the dish, it was easy enough to be persuaded that we might be watching some interaction of neurons. But when I looked closely at the termites, they didn't resemble hypothetical neurons. Nor did they act like computer simulations or robots. In fact, the closest thing to a robot here was Scott, who was using simple software to map the scrum of termites in the dish. Twenty-five clicks per frame; 120 frames per hour; a night's worth of termites sharing fluids.

By contrast, being a termite looks like a lot of fun. And that fun may be the whole point: perhaps there is a mad party-like vibe to termiteness. There is an old canard about termites building civilization without reason, but watching them, I wonder if there can be any civilization, or organization, without joy. Back in 1920, William Wheeler wondered something similar in King Wee-Wee's joke testimony. "Our ancestors did not start society because they thought they loved one another, but they loved one another because they were so sweet, and society supervened as a necessary and unforeseen by-product."

The role of joy in social organisms is not something we have a metric for, so it's not anything that modern biology entertains seriously. Robots and virtual termites have rules, but the rules of socialness— these urges and possibly even intentions—are unknowable to us. Watching this party, we find it hard to separate the building imperative (that possible stigmergy) from the termites' strange sticky social nature. Maybe they build the mound because it's fun to do it together. Maybe they transfer water because they're thirsty and moving the stuff around feels fun and necessary. And on this feeling of fun, perhaps, entire ecosystems are organized. Scott once wrote an utterly serious almost-koan about this: "A termite is not *is*, a termite is more *does*."

A termite is more *does*. Scott had showed me what happened when he dropped twenty termites in a dish. At first they wandered aimlessly, as if admiring a flower garden. Termites' togetherness is not merely an artifact of their evolution: what they are doing at any given moment is determined by how together they are. By the time there are forty termites in a dish, they start circling and then run faster and faster until they're thundering around the edges of the dish in a 240-legged herd. Perhaps they were sensing one another with their antennae, or were stimulated by the sounds of their running feet, but they carried on with jagged inertia. This seems like a trivial observation, but it's not: termite behavior is classically nonlinear—forty termites are not like two groups of twenty termites.

What this means is that termites have at least two fundamentally different ways of being termites: in the first they wander around at random, maximizing their exploration. In the second they are a focused force running toward a hole in the mound or digging a new foraging tunnel. As a system of work, the first phase allows them to explore all options without prejudice and the second allows them to change their environment rapidly. Scott said that sometimes a single termite, running in the opposite direction from the rest, can get the entire crowd to turn around and run counterclockwise.

Being a termite—never mind doing a termite—means being embedded in information that's mediated, reflected, tuned, and regulated by the mound. When Rupert got a sound engineer to put sensitive microphones in a mound, they discovered not the silence I feel near a mound but a roar. I put on the headphones to listen and immediately felt discombobulated, like when I'm in a subway station with horrible acoustics and trains are screeching through—only worse. My ears were overwhelmed by a racket of crackling, sparkling, ripping, and intermittent clapping. Behind this, like a timpani drum, there was a wavelike roar, which was occasionally washed by another, different roar that could be compared to a splash. None of the sounds made any sense to me. Rupert said the sticky noise seemed to be termite feet in the sticky mud. Just that detail caused me to reconsider my mental model of the termites as residents of Terminal Five at JFK; they actually live in a damp sewer. The overall roar might be stridulation, where the termites rub noisemakers on their legs together—something like scraping a violin bow. Termites signal one another by shaking their heads and vibrating back and forth—perhaps this accounted for the crackling noise. The very deep shoo-shoo sound and the splashing sound might be the sounds of air moving. The overall sense of resonance, according to Rupert, was due to the tuning of the airways in the mound itself. Whether the mound evolved for air or water, it's now part of a soundscape that makes sense to the termites. Just as the builders of early cathedrals tuned their acoustics for their music and their sense of celestial resonance, the termites have tuned theirs. Still, the noise was overwhelming. "Six hundred thousand legs are going to be loud," Rupert said.

If I were really a termite, this dense Jell-O of sound and tactile sensation would also have a deep, intense smell. Every colony has its own signature scent, an oily hydrocarbon molecule that covers the exoskeletons of all the termites in a colony. This scent allows termites to know immediately whether they're near a family member or an invader. Then there are other smells: pheromones, stomach contents, feces spread on the walls that repel invading fungi, and of course the smell of the big fungus itself. Any puff of fresh air, any strange-smelling termite, is immediately obvious, and everyone is roused to follow that smell or sound to its source.

Except when they aren't.

Scott has spent a few years trying to trick the termites, or as he describes it, "disrupt their perceptual chain." These tricks, or cognitive traps, attempt to prove the existence of the mound-as-semi-brain by demonstrating how to break it. For example, it's possible to stuff a lot of termites in a small space so that the tactile stimulation they get from one another is overwhelming. Then, when you attach a CO_2 canister to their petri dish and administer a clear signal to run, they will be so zonked out on togetherness that they fail to recognize the new reality. Instead of running they stay put.

Likewise, if you place termites in an area with no stimulation—

no air fluctuations whatsoever—they will soon congregate in little groups, stroking one another with their antennae. And when that air mixture changes, they will be too wrapped up in stroking and getting stroked to care. "My working model is that the high-intensity tactile feedback is so intense it takes over their whole cognitive world," says Scott.

It's a little harder to trick termites in their own mound, but it can be done. Normally, if you drill a hole into the mound, termites will wall off the end of the hole to stop the wind. Without any air movement or encouraging smells to stimulate them, the termites wander off to do something else. However, Scott used a piece of PVC pipe to trick the termites into thinking they had an ongoing disaster. The termites quickly plugged the hole in the end of the pipe, but the solid plastic shielded the freshly built walls from airflow, keeping the dirt moist and smelly, which stimulated the termites to keep building frantically, until they'd completely packed the pipe with mud balls. In this case, the pheromone—or the building scent or moisture or whatever it is—acts like a memory that won't fade, keeping the termites on frenetic, pointless overdrive.

The U.S. Army was paying for these experiments because they wanted to understand this "thought" process, whatever it was. Scott's time tricking termites allowed him to mull bigger questions. "Why do you need a brain for cognition to begin with? There's a metaphor of a brain as a computer, but I think cognition comes from the drive to homeostasis. Why does this system arise? Partly it's brain genetics, but partly it's because the biological system builds itself. The brain starts to make sense as an ecosystem. And the computation that the brain does is an emergent property, it arises between the cells."

He was still clicking and occasionally stopping to mutter about who was doing what, and how long they'd been drinking from each other. Scientists more or less agree that termites have some sort of collective intelligence—they "think" somehow as a swarm—but there isn't much agreement on whether they have individual intelligence. "For a long time the dogma has been that they can't learn, but who knows?"

As Scott's fluorescein studies went on, he found the termites sharing both from the mouth and from the anus, with all sorts of flank stroking to get a little trophallactic action. But sometimes, one termite would stroke while another snuck in to grab the goods. It wasn't an either-or situation about altruism vs. selfishness. It was all part of a grander social spectrum that kept the microscopic bucket brigade of water and nutrition running through the mound. Scott was in an excellent mood, and poo-stealing delighted him. He laughed, "All this horseshit kumbaya stuff about termite mounds!" Eugene was still around, and he snorted in reaction.

Rupert and his men had left, so it was just Eugene, Scott, and Vincent Mughonghora, a Namibian grad student with a wry sense of humor. With the daily rains the fungus had taken over the lab with a persistent, pervasive funk that I couldn't seem to shake. My clothes were perpetually damp. My sleeping bag was a tacky cocoon of must. I spent an afternoon pounding fungus comb until it was dust, which made my nose run, though it did reveal that the termites had moved the fluorescein dye all through the fungus, if not the mound.

Aggravating the general fungal air were the three *Termitomyces* mushrooms that had melted into a bubbly pile of brown slime just outside the open door. We had picked them from the top of a termite mound, but never got around to cooking them. This was ridiculous. The biggest mushrooms were worth twenty dollars, and one of Vincent's aunts cared for a termite mound in her yard because it supplied her with these tasty treats. Vincent made a few comments and went back to sending texts on his phone.

I felt trapped by the rain and wondered aloud if I'd be able to get out of the muddy farm road in my rental car. Was the fungus influencing *me*? It was hard to imagine any reality beyond this one of termites and ideas.

I was packing to leave when Eugene made a joke about crocodiles. He was building a black-light arena for the termites in the pantry, next to the onion bag. As he taped over the pantry windows he shouted, "Termites, you little green men from Mars! Now you'll spill your secrets." That got him thinking about life here under South African rule and a paramilitary police force that was so brutal the joke was that they'd beat a lizard until it confessed to being a crocodile.

With that, I was shaken out of the alternate reality of the farm and fungus. Reporting had been a cognitive trap where I faced inward, and so I had been missing all the cues: the road built for tanks, Hindenburg Street, even, it turned out, the origins of the farm we were staying in.

At the turn of the last century, Namibia was a German colony. Between 1904 and 1908, the German government followed a policy of "extermination" and forced labor in concentration camps, killing as many as 70 percent of the Herero people, the largest ethnic group at the time, as well as many thousands of other Namibians. The very land we were on had been taken from families of the Himbe and Herero ethnic groups in 1907. Many scholars consider Namibia's mass killings to be the first genocide of the twentieth century.

After World War I the country, then called South West Africa, was made a protectorate of South Africa, which enforced a variation of its own apartheid system. Between 1966 and 1988, local guerrillas called the South West Africa People's Organization, or SWAPO, fought the regime, and both sides were involved in the war in Angola. Namibia finally became independent in 1990 and the former SWAPO guerrillas were elected to power, forming a socialist government that has managed the country's uranium, diamonds, and dazzling crowds of zebras, elephants, rhinos, and lions ever since. Engrossed by termites, I had missed almost all of this.

The next day I got back in the car, drove around the muddy holes, and went past Hindenburg Street and down the straight road toward Windhoek. It all looked different. The mounds that had greeted me on the drive up now nodded in my direction like old friends. I didn't have a neat tale of superorganisms and robots on Mars. Instead, I'd seen something shifty, almost Hegelian: a conversation between failure and survival, sun and rain, being and doing, one and many, metaphor and theory, robots and joy, mud and cognition.

SEVEN

A BLACK BOX
WITH SIX LEGS

I**N THE SPRING OF 2012,** I went back to Namibia. As part of his goal of working with programmers to figure out how and why termites build, Scott had invited a team of roboticists led by the Harvard professor Radhika Nagpal to the farm. When he told me, I decided to pay my own way to watch them work, telling myself that it would be a vacation. What could be more rejuvenating than seeing the termites again, this time through the roboticists' eyes, and taking notes?

Scott's lab was taken over by technology. In the spot where I had pulverized fungus sat a laser scanner draped with a hand-sewn black cover that made it look like a scale model of Mecca. The back room where I had slept had been hung with cameras and lights trained on Plexiglas obstacle courses designed for termites. The windows were darkened with a piece of purple zebra-print cloth.

Radhika was slight and animated, wearing a large orange fleece and a pair of navy-blue polka-dotted rubber boots that anchored her to the ground. When she was excited by an idea, she moved her hands and began to sway in the big boots, causing the black ringlets in her hair to bounce in a wave action of enthusiasm. At forty-one, she had just gotten tenure at Harvard. We had hardly finished pleasantries before the conversation began to ricochet from small ideas to big ones; slime mold to skyscrapers. When she was really excited, she leaned back and laughed, as though the force of the idea had blown her backward.

By contrast, the members of her team were solemn, biting their lips as they typed swiftly on their many laptops. They included a physicist in his thirties, as well as a computer science postdoc and an electromechanical engineer—both in their twenties.

Radhika had started studying computer science as an undergrad at MIT, moved into "hard-core" problems like circuit design at Bell Labs, and crossed over into biology during her doctorate at MIT, when she tried to imitate the way individual cells act in concert as "amorphous" computers. After that she moved into robotics. Now she is part of big teams working on National Science Foundation grants that have a futuristic bravado: "Programmable Second Skin to Re-educate Injured Nervous Systems" and "RoboBees: A Convergence of Body, Brain and Colony."

Meeting Radhika explained something I'd noticed in the videos on the website for her lab at Harvard's Wyss Institute: all of her team's robots were funny. They built structures by piling blocks, spraying foam, and flinging toothpicks, but each one had a careless, comic self-awareness—like a bored jock in a high school cafeteria.

The most famous robot at the Wyss Institute, though, was the Robo-Bee, made by a team led by Rob Wood. Paper-clip size, with the spare esthetic of industrial jewelry, the RoboBee can take off, fly, and more or less land; but it's trying too hard to be a real bee to crack a joke. During its short non-life the RoboBee has become a working symbol of the potential of swarms of cheap, insect-like robots. While

Radhika's group didn't work on the design of the RoboBee, they were involved with thinking about swarming.

The RoboBee was seductive; I couldn't help but imagine what I'd do with a bunch of such things. Neither could anyone else: it had been pitched as the future of search and rescue, environmental protection, traffic control, agriculture, and of course the military, which hoped to use swarms of tiny autonomous vehicles to go to places too dangerous for human soldiers.* In a 2010 video an enthusiastic assistant professor made a presentation about using RoboBees to replace real bees to pollinate flowers, which seemed far-fetched considering how expensive they are now. Still, the idea was alluring, and an audience member asked how he'd keep birds from eating his robots. Enticed by his own story, the professor said he'd probably need to hire people to walk around the fields picking up robots that had crashed to the ground.

The first time I talked to Radhika, on the phone, I wanted to talk about swarming RoboBees and the like, but she was reflexively dismissive, complaining that they can only be maintained by PhDs and are merely "a shadow of what exists in biology." She wanted to talk about watching humans sightseeing along the Charles River. "Why is it that all of us are walking around and then suddenly everyone is taking pictures? Something gets set in motion and there's an implicit chain of events." The technical term for that chain reaction of one human clicking a camera leading to the rest of us taking photos is "propagation of local effects." She went on to explain that the particular problem in swarm intelligence that she is trying to solve is called "global-to-local," which means she needs to understand how the global response of the crowd forming starts with the specific (local) cues that make a person start to click. Global-to-local is the roboticists' name for the way that termites build—coordinating thousands of individuals with a simple ruleset for every local termite but apparently no overarching plan as that global mound takes shape.

---

*The RoboBee was funded entirely by a large multiyear grant from the National Science Foundation. Robert Wood, the leader of the project, had previously received funding from DARPA.

Understanding the rules of global-to-local would give researchers a guide to complexity in the form of a set of simple instructions—an algorithm. With this algorithm they could build robots that could themselves build buildings in remote places—like Mars, but also your backyard—without central instructions. That algorithm could also be adapted to software so that computers could coordinate themselves to fix software, or broken networks, much the way termites fix holes in the mound. These swarms, whether real or virtual, could potentially have intelligence without reason: as clouds of drones they could find trouble, and as software they could solve problems. The Robo-Termite didn't have the flash of the RoboBee, but it was a very powerful idea.

I DIDN'T RECOGNIZE the experiments in the lab. For example, there were petri dishes on the table, but each one was filled with pink, blue, or green plaster that had been painted with a thin layer of dirt. They didn't have the improvised weirdness of Scott's experiments. In fact, they were quite pretty, and had been molded into three-dimensional patterns: a raised daisy shape, concentric rings arranged like tiny amphitheaters, and wedges you might find on the sole of a running shoe.

"Those are chirps," said Nils Napp, the computer science postdoc, pointing at the wedges. Nils has a boyish freckled face and is deliberate in all things. He said the word "chirp" with a little trill, as if chirping himself. A chirp? I was blank. Chirps are what engineers use to test black boxes, he said. Before we could talk termites we had to build a bridge of language between us.

To engineers a black box is a machine (or situation) where the contents and capabilities of the interior of the box are unknown but the inputs and outputs are known. For example, if you put a tone from a guitar into an electronic system and get a distorted crunchy sound out the other side, you know that your "black box" is the kind of box you'd attach to a distortion pedal on an electric guitar. One way engineers test black boxes is to feed a known quantity like a sine wave in to see how the black box changes it. Thus, engineers feed "chirps" of various changing frequencies to black boxes and analyze what comes out.

What I am looking at in the bottom of this petri dish—the series of ridges that resemble the bottom of a track shoe—is a physical chirp, a series of sine waves carefully painted in soil that will be presented to the termites for modification. The plaster surfaces were brightly colored to make the places where the termites had removed soil easy to see.

This idea struck me as incredibly strange. Turning an electronic pulse into a termite playground seemed to be mixing metaphorical valences—as weird and inappropriate as handing a mechanic a photograph of a screwdriver instead of a real one. But Nils hoped to push the idea further by putting the petri dishes under the laser scanner and continuously scanning them. Then he wanted to digitally remove the termites from the 3-D models to see what had happened to the dirt. Termites, he said, were "noise," a distraction. In a perfect world he would reconstruct what termites do by ignoring them entirely. This was an entirely different way of thinking about a termite.

I sat down near the wall and watched Nils and his fellow researcher Kirstin Petersen work without talking. Kirstin is small, with a blonde ponytail and clear blue eyes that are usually narrowed with intention. In this group of very precise people, she distinguished herself by being both more exacting and more retiring. An electromechanical engineer from Denmark, she'd previously worked on robots for outer space. Now she was designing and programming the lab's termite-based robots, called TERMES. She had built the laser scanner from mail-order parts and written the software to run it. When Nils had an idea for the chirps, she 3-D printed the molds to produce them. She'd also designed the lab's T-shirt and was preparing termite samples for electron microscopy. Her tough competence was the operating system the rest of the lab ran on. But as the lab's junior member, she had developed a set of self-effacing tics—a giggle, a rapid mutter, a lock of blonde hair that always fell out of her ponytail and into her eyes. As

Nils talked, Kirstin muttered: "We did a beautiful one last year but the camera broke."

The termites did not cooperate. After placing twenty termites in a dish, Kirstin and Nils carried them back to the darkroom and put them under the bright lights and cameras for recording. Some groups of termites twiddled their antennae for the forty minutes they were under the lights. When their dish had been discarded, they'd work up a frenzy of destruction, pulling apart the shapes and remodeling them—but away from the camera's eye. Scientifically speaking, the frenzy had never happened. The chirps attracted some building, but then the mud balls all rolled into the valleys, so it was unclear whether the termites were placing balls on the low part of the chirps or whether they were placing them on the high parts and gravity was rolling the balls to the lowest places. For all we knew, the termites were playing with the chirps to see how the black box worked.

I MISSED SCOTT, who had been delighted by the arrival of Radhika's team. His big questions were starting to draw a crowd. In addition to Rupert, he'd recently attracted some physicists from Harvard who were going to investigate how air flowed in the mounds, an Indian neuroscientist, and a new postdoc. He was no longer working alone with his big ideas; he had even slightly unhunched his back.

But a day or two after I arrived Scott got word that a family member was in the hospital, and he had to leave. Berry Pinshow, a passionate South African–Israeli physiologist who had come to Namibia to help Scott make educational videos for local biology students, drove Scott to the airport, and I went along. Berry had met Scott in the 1970s when Scott was on a mission to borrow a frozen alligator. On the way home Berry and I were both hit by sadness and empathy for Scott's years of work far from his family. Though Berry was normally boisterous, we both worried. After that ride we became friends and allies. He was a fine and devoted observer of birds, and once when I described a strange one—I didn't know what I'd seen, but its wings did not appear to be hinged at the usual spot—he stared at me for a moment as if I weren't there. Then he moved his arms exactly like the bird's wings and told me I'd seen a crowned plover. Berry took a dim view of the roboticists: they weren't "thinking like termites," he said.

RADHIKA CAME TO THE LAB and stood in the middle of the room and said nothing for about ten minutes. She swayed silently in her polka-dotted boots, and I couldn't tell whether she was pleased or displeased or thinking about something else entirely. She exercised a subtle charisma: rather than infusing the team with an oversized scientific superego, an angel-devil who sits on their shoulders making wise-ass remarks, Radhika seemed to have colonized a quiet, cerebral space within their brains. The roboticists habitually self-checked their internal circuits and assumptions. Nils and Kirstin were self-contained in the extreme, furrowing their brows as their darting fingers struck the keys on their laptops.

Later Radhika explained that she didn't want them to be dependent on her for leadership, so she stood back while they tried to solve problems, which sometimes used up extra time. It seemed fitting that she'd try to achieve swarming, leaderless, global-to-local productivity among the humans in her lab.

She began asking questions circuitously, noting that the termites working in the chirps were giving information only about removing mud balls, not about stacking them, so the team members weren't collecting data on stigmergy. "It's a nice technique for seeing what they pick up, but I'd like to know where they put the balls down," she said.

"I'm excited that they dug anything," said Nils. Their conversation lurched along, but it never felt quite like a natural conversation, more like two separate ones that had been spliced together.

As they talked I realized that they were desperate for a different kind of data from Scott's. They didn't want termite tendencies, or stories about what the bugs liked and didn't like and why they'd evolved the way they did. They wanted to record the termites doing the same tasks over and over again and then use those histories to make predictions about the way that termites behave. Anything less than this was failure.

THE FINAL MEMBER of Radhika's team, the physicist Justin Werfel, rarely came out of the converted bedroom with the camera equipment. His official position was staff scientist, an undefined rank well above postdoc. He had exuberantly curly hair and he often clenched his jaw as though preventing himself from saying something. His aura of concentration and control was so strong that I wasn't sure if he would talk to me when I interrupted him in the midst of rigging some LED lights and a motorcycle battery over one of his Plexiglas bridges.

When I asked him how he got into termites, he dropped the equipment he was working with, backed out of the dark into the hallway where it was light, and fixed me with his grayish-green eyes.

"Termites are my whole career," he said. "But by termites I mean artificial termite colonies." These computerized termite swarms are part of a search for scientific truth that started in high school, when he fell in love with physics as a reductionist way of understanding the world. He came to realize that physics was obsessed with math problems, so he moved on to neuroscience, which he felt asked bigger, wider questions. Over time, and with different mentors, he turned to studying how consciousness arises and how cancers are organized. Eventually, he had the sickening feeling that these mysteries would not be explained in his lifetime. He read the first complexity paper on termites in 1997 and started programming termite models in computers in 2003.

Justin spoke quickly and intensely, his words coming from a cool box of rationality far behind him and translated over a long thin wire to me. This feeling of distance made me return his gaze with greater intensity until we were locked in a feedback loop of intense eye contact, despite my growing panic that I didn't understand anything he was saying. Here is what he said about observing termites: "We always start by saying we're inspired by termites, but in fact it's not really known what they do. What program are they running? Can we get a stochastic model of a stateless automaton that has no memory but reacts to what it encounters? Can we connect the global to the local?" And with a self-conscious giggle, he broke the gaze and spread his fingers wide, in a gesture of helplessness. "It may not be possible."

I had to go outside to translate what he was talking about. While you or I might look at a termite in a dish and recognize a simple animal, he wants to describe its behavior as that of a simple machine with no memory and a limited number of actions. It can turn left or right or pick up a ball of mud or put it down according to cues it gets from its environment. In any situation some actions are more likely than others. If the roboticists could capture the data from these interactions over many thousands of iterations, they could build a model of an idealized termite. The termite would run stochastically—by chance—and with every roll of the dice the termite would take an action. When these actions were combined across hundreds of autonomous computerized termites, something new would emerge—hopefully a mound.

Here was a case where I should have believed Berry's complaint that the roboticists weren't "thinking like termites," which I had dismissed as cranky.

The idea that termites can be thought of as robots—or, more generally, that all nonhuman animals are robots—goes back to René Descartes, who died in 1650. It was Descartes who described animals as "soulless automata," in contrast to humans, who had souls, minds, and intentions. Descartes drew this distinction because serious thinkers of the time needed to separate the natural from the supernatural, and they needed to study animals in ways that did not step on the Church's toes (hence only humans had souls).

The robots that Descartes wrote about were not like ours today. He probably saw a famous tableau of bronze automata at the Royal Gardens in Saint-Germain-en-Laye, with a naked mechanical Neptune shaking his trident from a chariot pulled by seahorses, surrounded by horn-playing mermen. The whole thing was powered by flowing water.

At a time long before movies, when clocks were cutting-edge, the Gardens must have been amazing to watch. By comparing your average dog to these extravagant mechanical amusements, Descartes was implying that God was a great engineer. That rational, engineer's mind was something Descartes wanted to capture and apply to the world, tracing out great chains of cause and effect. We continue to look at termites (and biology in general) through Descartes's eyes: Scott's dogged process of elimination (which is called "Cartesian doubt") is one legacy from him; the graphs we use to map two factors at once, Cartesian coordinates, are another.

Still, can termites be considered "stateless automata"—memoryless identical machines that only react? Had I, under the influence of both Eugenes Marais, imagined all the crazy social complexity I'd seen on my last trip? The researchers were so bright. So jittery. I wanted them to will a stochastic termite truth into revealing itself and I hoped to see the world anew in their data.

KIRSTIN HANDED ME a bucket of fresh mud and termites. I got a small paintbrush and began to sort worker termites into groups of fifty As soon as I dipped my paintbrush into the termites I felt good.

Bug vision—watching termites crawl and shake until the mud itself seems alive—is fun. I got some termites to grab the brush, then blew on them until they dropped into the petri dish. My brush was a little chairlift to quantification. I did that same motion about thirty times, getting one or two termites at a go. Then I started on the next dish. Watching termites requires that you turn your internal excitement meter down to just about zero, which means that even the slightest variation in their behavior or patterns—say, some clustering in a small cul-de-sac of mud, or a few shaking their ridiculously large Shar-Pei–like heads—seems wildly stimulating and fulfilling. I suppose this is why some people meditate, to observe minutiae in a sort of fugue state. It's clearly one reason biologists are attracted to observation—that ability to stop being a giant human and attend to the small scale and fast time signature of the other. What kept me coming back to termites—paying for tickets to these faraway places—was that watching them gave me a satisfying signal and hardly any noise.

Radhika stood over one of the petri dishes, watching the termites hug the edges of the flower-shaped patterns. She tried to imitate their choices, holding her arm out close to the wall as if it were her antenna. Because they have no eyes, termites will run an antenna along a wall to track where they are. With ants, when the wall disappears they'll turn only if the angle is less than 60 degrees. Over 60 degrees they'll turn around and run back. Under 60 and they'll make a choice to turn right or left, but some have what's called a "directional bias" of tending to turn left when they lose the wall. Ants, then, can be tracked in a way that's relevant for developing computer models.

The petri dishes with flower shapes gave termites curves that faced in and curves that faced out in the hope of eliciting this kind of behavior. Radhika paced around the table as if she were a termite in the flower-shaped dish, following the wall with a bias toward the left. Watching Radhika was as close as I got to understanding whatever it's like to be a termite on this trip.

One of Berry's accurate observations was that the roboticists preferred eating peanut butter alone at their computers to joining the communal evening meal. Berry was incensed by their lack of interest. It wasn't just that they were eating peanut butter when they could be eating an oryx steak; they were living life wrong! And not knowing how to live was a crime, in Berry's book, right up with failing to note a visit from a lilac-breasted roller, which is a bird as spectacular as its name suggests.

I was on Berry's side. We went shopping to prepare blowout meals with grilled game steaks, local sausages, and stuffed gem squash. One night when Berry called the roboticists to dinner they didn't look up from their screens. He stood in their midst, comically clutching his hands to his chest and calling them "my sweeties" as though they were his grandchildren. They left their work with reluctance.

At dinner Radhika mentioned a professor who said that termites' ability to work together should be applied to politics. "I don't want to save the world. I want to make robots," she said. When it was fully dark, Berry pointed out that a family of very small owls was living in the tree above us.

After dinner, Berry lay on the ground blowing cigarette smoke into scorpion burrows while the researchers tracked the smoke's meandering exit with a cheap green laser beam. Berry had a theory that scorpion burrows, like termite mounds, manipulate airflow, and he hoped to expand the theory by observing the escape of the smoke. The green laser doing its goofy psychedelic dance in the smoke under the stars was nothing they'd ever trained for, but Kirstin and Nils pitched in, balking only when Berry suggested that Kirstin lick the camera's batteries to get a better connection.

Biology was starting to unsettle the roboticists. While we washed the dishes, Radhika said she was getting flashbacks to early in her career, when she first crossed over from computer science. "I had forgotten that feeling of voodoo when you're working with biology," she said. Compared to engineers, she said, biologists are like cooks or magicians who do experiments that no one else can duplicate. During the year she spent in biology labs, it wasn't just the methods that perplexed her, it was that the sensitivity of biological systems was so far from the linear sensibility of machines.

While doing biology, she had begun to believe in a sort of witchcraft. "I started to make bargains in the lab." When she was trying to do an experiment that required her to cut frogs' eggs in half, very quickly and precisely, she was told that the best thing to use was a blond baby's hair because it was the thinnest, strongest material. Part tool, part good luck amulet, and a side commentary on the Anglocentrism of biology. "Blond!" she exclaimed, laughing.

The experiment, when it worked, demonstrated something bizarre about how organisms build and scale themselves. If you cut a frog egg in half at just the right moment, you will end up with two half-size frogs, suggesting that there's some specific instruction about the scale of the frog in its embryo at that specific moment, a crucial bit of local-to-global code. But Radhika's experiment failed. One of the biologists said: "Maybe you didn't have a robust frog." The very idea of a robust frog or a non-robust frog does not exist in computer science. How does anyone spend a career doing experiments that can't be replicated? The endless weirdness of the humans—rather than the frogs' eggs—cracked her up.

Radhika used humor and frustration—both symptoms of the same problem—for forward leverage on these biological mysteries. With each failure, she reinvented the question so it was funnier.

One evening she mentioned that when she was young, her parents moved from the United States back to Amritsar, India, which later ended up under government siege following a bloody raid on the Sikh Golden Temple in 1984. During her teens in Amritsar there were years of curfews when she rarely left the house alone. Her humor was a philosophy and a tool, a kind of sword she'd adapted to use in the world.

EIGHT

WAITING FOR CARNOT

ONE NIGHT I HAD a vivid termite dream: I saw a string of sausages moving across the ground: my picnic had been stolen by insects. I watched them parade my sausages all through town to a box they'd also stolen, where they were hiding the whole town's sausages. I woke up feeling bothered and ripped off.

When I told Radhika about the dream, she said sympathetically, "Oh, collective transport," as though she had that dream all the time. Whatever frustrations I felt were transient: I was merely vacationing in the roboticists' nightmare. Finding the termites' modus operandi— and that of other self-organizing systems—was a lifelong project, a vision quest, not a project with a completion date. Radhika and Justin had been living this way for years.

Justin explained later, a little ruefully: "There's a joke in Complex

Systems that we're all just Waiting for Carnot." This is a reference to Beckett's absurdist tragicomic play, *Waiting for Godot*—in which two guys wait for Godot, who never arrives. Carnot is the French engineer and physicist who, in the 1820s, first observed and correctly described how heat energy is converted to work, laying the foundation for the laws of thermodynamics. Before Carnot there were observations of heat's behavior, but his theories transformed early steam engines and influenced the design of gasoline engines, power plants, and the foundations of our modern world.

The field of complex systems is still in the stage of gathering insights into biology while waiting for someone to appear with a unifying theory. Come up with a viable theory for the way termites build and it could change the way computer networks run, how wars are fought, and how disasters get responded to. The emergent equivalent of thermodynamics could upend the world.

But the reality, like the play, is absurdist. Fifteen years earlier Justin had gone to study consciousness, only to wake up in a Beckett play full of termites that might not end in his lifetime. He was adapting well, considering.

One evening Radhika and Justin huddled in the back room trying to figure out how to reduce the termite's movements to probabilities. They talked easily in a kind of fast code. When I entered, they switched to language I could understand. They had started wondering if the termites had memories. Maybe termites acted differently when they're hungry or thirsty—conditions that could function as a type of memory, because an empty stomach is a marker that food has not been eaten.

"The probability of a reaction is never zero, and it's never one hundred percent," Radhika explained. "Maybe what we're seeing as probability is actually personality or some other data."

To get data, the roboticists had to wrestle with a big question: Can a termite's behavior be described by an automaton? They began to question every one of their original assumptions. They wondered if the termites were really stochastic. Could they have complex memories and be reacting to something twenty steps before? Radhika mused, "Should we worry that we're just modeling our own assumptions? Are the termites random, noisy, or something else?" The very concept of the black box might be a kind of cognitive trap that was preventing the scientists from seeing that the termites were, at some level, doing.

Justin had four petri dishes under the camera at once, with one termite in each dish. He looked in at them and saw that only one of the four termites was walking around. "This is terrible," he said. Radhika tried to sum it up statistically: "I think it's likely that just one in four moves. I think that's reality."

Radhika and Justin stood in the center of the room and leaned back slightly at the same angle, training their eyes intently on the experiments that were lined up on the floor in the cupboards, beneath the cameras.

I had crouched by a table to take notes, and I noticed a dish where the termites had been slacking under the video camera. Now that they'd been discarded to the side, the termites were stacking up balls vigorously, like they had a project in mind. Justin and Radhika stared at them sadly. When they looked back at their current experiment, three out of the four termites were running around in their dishes. The odds had just reversed.

Radhika snapped out of the termite puzzle and looked at me. "People don't study termites because they're hard," she said, "but I'm happy discovering anything about global-to-local that increases the chances of awesomeness."

In the background, Kirstin was doing a series of very simple experiments in petri dishes. She filled one-half of each dish with dirt that had previously been built by termites and put fresh unbuilt dirt in the other half. Then she put termites in the dishes and placed them under a camera. She was trying to gather data for one of Scott's big questions about whether there was a "pheromone" on the dirt that would influence the termites. The experiment, however, was an afterthought, and it didn't seem like a big deal. The dishes themselves were not at all pretty; they certainly gave no sense of the secrets they would reveal a few years in the future.

I FELT SORRY for the roboticists and the messy progress of their work, and that was made worse by the fact that once I got to know them I liked them. I started withdrawing to read.

One of the books I read was about Namibian history, which led me to dig more deeply into the subject. I came to understand the central role of the machine gun in making the country I was now in. In 1900, after two decades of occupation, Germany's attempts to make Namibia a colony were floundering: the Germans were losing territory, their soldiers were sick, and the people back home didn't want to keep paying for the colony. By 1904, the Herero, who were good fighters and excellent strategists, controlled most of the countryside and many of the cattle that had belonged to German settlers.

Maxim guns were then a new weapon, used by colonial powers only in Africa, the Middle East, and Tibet. In 1898 the British used the guns to kill ten thousand men in Omdurman, Sudan, while losing only forty-seven of their own. They won completely—and easily—but a European witness described it with distaste as "not a battle but an execution." The European officer class felt the weapon could not be used in European wars requiring honor, chivalry, and victory by "fair fight."

On August 11, 1904, the German general Lothar Von Trotha, newly arrived in Namibia, placed troops carrying fourteen machine guns around an encampment of about six thousand Herero fighters and as many as forty thousand women and children, and gave an order to "annihilate." Over the course of the daylong battle, people escaped into the desert, which was unguarded. Many later died of thirst and starvation there, chased by "Cleansing Patrols." Von Trotha vowed the Hereros' "extermination" with this statement: "I shall destroy the rebellious tribes by shedding rivers of blood and money."

A Herero mourning song described the deadly exodus as beads falling off a woman's neck, with no one to gather them up. The beads were symbolic of the community, but it's said that real beads can still be found in the desert to the east of the Waterberg Plateau, which is not far from Otjiwarango.

Thousands of Namibians were later sent to concentration camps, where the majority died. Many of their skulls were sent back to Germany as support for racist theories of white supremacy. Repatriation of the skulls began only in 2011.

I wondered about the ideology of the machine gun: If it was morally unsuited to Europe, what made it an acceptable tool for "extermination" in Namibia? In Germany, as in many other colonial powers, Africans were dehumanized and seen as "inferior" to whites. This fit with an utterly perverted interpretation of Darwin's work that was used to argue that Africans would "naturally" become extinct, like dinosaurs. I found this disquieting mingling of sentimentality and excuses for genocide in a book called *Savage Africa*, written by the British traveler William Winwood Reade in 1864. "They [Africans] may possibly become exterminated. We must learn to look on this result with composure. It illustrates the beneficent law of Nature, that the weak must be devoured by the strong," he wrote. Reade expected that Europe would divide up Africa, drain it, cultivate it, and turn it into another Europe, and "young ladies on camp-stools under palm-trees will read with tears 'The Last of the Negroes,' and the Niger will become as romantic a river as the Rhine."

Europeans' misguided ideas about the machine gun prevented them from grasping the power of the technology itself. Because the machine gun was seen as fit only for colonial wars, European officers did not plan for it to be used on their own troops. They "clung on to their old beliefs in the centrality of man and the decisiveness of personal courage and individual endeavor" on the battlefield, according to the historian John Ellis in *The Social History of the Machine Gun*. Caught in their own trap, these elites believed that such weapons would never be used in Europe because their wars, as they saw it, ran on honor and chivalry rather than rivers of blood. And so most of Europe was unprepared for the machine guns, the poison gases, and the bombs of World War I.

Whenever I looked up from this reading, there were pans full of mud and buckets of termites all around me. What if humans did gain the powers of termites? What then?

THE TERMITES STAYED inscrutable, but the scientists were changing. Kirstin asked me why dinner was important to Berry. I guessed it was because biologists in the field need to trust one another with data, methods, and even their health; dinner bonds them together. One entomologist had told me that the emotional closeness of biology field-work reminded him of when he was a member of the cast of a musical. Kirstin was silent and then said that before coming to Namibia it was hard to work with Nils because they didn't trust each other. It wasn't that they suspected deceit; it was that they couldn't be sure they were talking about the same things or concerned with the same problems. Over the past week they'd been staying up late, drinking beer, and trying to fix the laser scanner. Now they trusted each other, she said, and were getting more work done.

Near the end of the trip, Justin stood in the back room grabbing his hair with his fists. He'd been repeating his tunnel experiment for days, but it had suddenly gone weird. The tunnels were Justin's solution to the randomness of termites. He intended to give the bugs just one thing to do and one direction to do it in, so the tunnels were constructed of Plexiglas, and just wide enough for two termites. The tunnels connected two chambers: in one there was dirt from a termite mound and in the other the dirt was fresh. The dirt from each side extended exactly halfway into the tunnel, and he intended to test whether the termites acted differently on the mound soil, which had termite saliva and possibly cement pheromones in it, than on the fresh soil. It was actually quite similar to Kirstin's experiment in the petri dishes, but Justin was hoping to create a controlled experiment, limiting the variables so he could gather data on the termites' behavior. But this time all the termites walked into the tunnel, stopped halfway, and turned around to dig furiously in the fresh dirt in the tunnel. Rather than passing one another neatly in 2-D space, they consolidated into a termite ball, a whirling 3-D dirt frenzy. No data could be obtained from this ball.

I'm going to step out of time as the narrator to say that in fact, Justin was looking at something big, possibly even something huge, but it would take the team years to understand it. In order to see it the researchers had to de-rationalize themselves. They had to learn to see and hear what the termites were doing: what the original Eugène Marais had called "learning a new alphabet." This would be hard for people used to working with machines. And it was especially hard for people who so strongly distrusted intuition. The mound and those termites—stochastic or not—would remain an enigma for now.

PART III

NINE

THE SECOND TERMITE SAFARI

NEVADA

JUST AS HUMANS DREAM of making robots to imitate the way termites build, we also want to copy the way they digest wood and grass, to make replacements for gasoline. As far back as 1881, scientists were describing the bugs as "a powerful mill." Now we wishfully call them "bioreactors." One paper described the segmented termite gut as a veritable factory: an enzymatic reaction chamber followed by a permeation filter, product recovery, and an anaerobic digester converting polymers to microbial products. Naming the things we're studying as if they're the things we hope to create is a mixture of sympathetic magic and salesmanship—but, like calling termites robots, it contains contradictions and paradoxes.

In October 2009, I was sitting at my desk thinking about oil when I got another email titled "Termite Safari?" from Phil Hugenholtz, the microbial geneticist who had worked on the first effort to find all the genes in termite guts, and who was my host on the previous year's trip to Arizona. Now he was offering a safari in Nevada, once again in search of genes that would contribute to the project of making grassoline. I went, paying my own way, rationalizing that I could do a conference call for work while hunting for termites.

After my time in Namibia, I was eager to tour the bizarre world of termite guts. If the termite mound is a composite animal composed of millions of self-organizing termites, the termite itself is another shell company for a consortium of five hundred species of symbiotic microbes, all cooperating to digest wood for the mutual benefit of the Many. Even better, some of these microbes are *themselves* conglomerations of several creatures acting as one. Perhaps Phil could lead me deep enough into this matryoshka world of minute cooperating souls to explore who is "in charge"—the superorganism, the termite, the microbe, or the microbe's microbe—and (recalling Scott's question about whether the fungus controlled the termites) just what "in charge" might mean.

I found Phil on a bench at the San Francisco airport, dressed like a student in a T-shirt, a hoodie, and sneakers with a heavy laptop bag. He was about forty, angular, disheveled, and frequently in motion. Now he was jangling over something on his laptop screen—a grainy black-and-white image of a thing, maybe a worm, wiggling through a wall. In his Australian accent, he exclaimed "Welcome to Science!"—as much to the image as a greeting to me.

Phil always reminds me of a member of the art-rock band Devo. Partly it's that he looks the part with his narrow face, square glasses, and the sense that he is always slightly out of place. But he is far too deep in his work to actually wear a flowerpot on his head; what I was really picking up on is that he is a professional surrealist, navigating huge quantities of data in search of meaning.

The screw-shaped worm on his computer screen was actually two spirochetes—long spiraling bacteria—wrapped around each other somewhere inside a termite's gut, rendered by a 3-D scanning electron microscope. The "wall" was probably the wall of a cell within a tiny piece of grass. In grass and trees, these walls are made of complex sugar structures, called cellulose and lignin, that are very hard to break down—which explains why wood makes such sturdy houses. Phil suspected the spirochetes had some kind of special enzyme capable of cutting the wall. If the lab could find these cutting enzymes and identify their genes, they might be helpful for the greater project of making grassoline. The stress of this mission, and the failure to fulfill it, weighed on Phil, but he said it as a rueful joke: "We can send a man to the moon. You'd think we could crack a few sugar chains."

WHEN PHIL and thirty-eight other researchers first did genetic analysis of the Costa Rican termites' guts in 2007, they found 71 million base pairs, or twinned molecules of DNA, which they sorted into approximately eighty thousand genes, and among those—using computers—they identified 1,267 enzymes that might work to digest wood. Press releases suggested that once the termite's gut was decoded, we'd soon be inserting these codes into tame laboratory bacteria to produce enzymes and start digesting wood on a grand scale. Phil, at the government's Joint Genome Institute (JGI), was soon working at affiliated labs that were focused on the grassoline problem: a new government-supported lab called the Joint BioEnergy Institute (JBEI); Lawrence Berkeley National Laboratory (LBNL); and a joint venture between UC Berkeley, LBNL, the University of Illinois, and the oil company BP called the Energy Biosciences Institute (EBI), which aimed to develop sustainable energy sources and chemicals.

But the termite, it turned out, was a hard bug to crack. By the time I met Phil again at the airport in 2009, it had proven to be much more than an exceptionally elegant machine, a natural blueprint for a factory, or a source of code to "boot up" a bioreactor. Phil said the details of how the termite's crazy consortium of microbes accomplished wood eating were still a mystery, and they had proven difficult to re-create in the lab. "The joke is that by the time you're done you'll have a termite, and you might as well go and hook your car to a bunch of termites."

Now a grant from EBI was sending the team to Nevada to get termites with different food sources—grass vs. cow patties containing partially digested grass—to see if they could unravel the termites' strategies by comparing their guts.

Our flight to Reno was called. As Phil stood up, he realized he'd lost his boarding pass, and Anna, the lab's tech, immediately jumped in. Anna wore a baseball cap with her ponytail threaded through the back and she had a pragmatic daredevil force of personality. As Phil searched the pockets of his pants and his laptop bag, she quizzed him aggressively, even though she was the junior member of the lab. She was responsible for making sure we came home with termites properly frozen in labeled vials, and she took her job seriously; it wasn't going to work if Phil didn't have his boarding pass. The other member of the team was a postdoc named Shaomei He. Like Anna, Shaomei had threaded her ponytail through her baseball cap, but rather than throw herself into the action she stood back, observing closely with keen eyes. Shaomei found the boarding pass and we all got on the plane.

On a different scale, and in a different discipline, imitating the chemistry of the termite is a variation on what Radhika's team is looking for—the secrets of global-to-local, a Carnot-scale insight into how the parts relate to the whole, whether that means the microbes in the gut or the genes in a bioreactor's warm steel tank. Still, I was surprised to find that this quest, too, includes concepts like black boxes and frustration with the "voodoo" of biology, not to mention engineers and physicists. Even more than the robot project, the biofuel one aimed high: it didn't just hope to change the fuels we burn, but also the foods, medicines, and chemicals we use.

HERE'S WHAT WILL happen when termites finally get around to eating this book: one will use the clippers on the end of its mandibles to grab a mouthful about the size of a period. It'll push that into its mouth, which resembles a grinder, with its hand-like palps. From there the shredded paper will make its way into the gut, which is about an eighth of an inch long and the width of a hair.

The first stop in the gut is a gizzard, where the bite will be vigorously mashed with saliva containing enzymes to grab any free sugars, which are quickly absorbed by our termite. Next, this paper bite will journey through an alkaline tenderization chamber for a nice soak in the termite's version of drain cleaner. After that, depending upon which kind of termite it is, the bit of papier-mâché will proceed through an elaborate enteric valve—a gorgeous gatekeeper made of many little fingers brushing the particle into the cavernous nightclub of the hindgut, named "P3."

The inhabitants of the P3 are glorious and bizarre. Microbially speaking, they're a freak show. There are, first, as many as fourteen hundred different species of bacteria. They are a collection of little spheres and rods and spirals, jumbled together like a box of sticky old Christmas ornaments. These microbes release enzymes that can unzip the cellulose and hemicellulose in our paper particle, producing sugars. All around are masses of other microbes waiting to grab the sugars and process them into hydrogen and methane. Along the way they may synthesize some nitrogen compounds, too.

This microbe community looks chaotic to our eyes, but in fact it meticulously arranges itself in neighborhoods where sympathetic creatures can eat one another's garbage. Those who are the most friendly with oxygen sit on the edges of the gut, while those who can tolerate none hang out in the middle. All termites have bacteria; but some so-called higher termites, like the fungus-growing *Macrotermes* of Namibia, have only bacteria.

By contrast, the guts of so-called lower termites host bacteria as well as exotic creatures called "protists"—single-celled organisms that are neither animal nor plant nor fungus. Protists are relatively huge and quite weird. If you were a piece of paper the size of a bacterium, say, and just entering the termite's third gut, you would be greeted by a giant swirling thing, three hundred times your size, approaching like a cruise ship coming in to a dock, so big you wouldn't have any idea how big it really was. That would be *Trichonympha*, the most common of the termite protists. It has a smooth, round cap, like the tip of a badminton birdie, and an enormous whirling hairball, made of thousands of flagella over its barrel-shaped body. Opposite the tip, buried under all the waving flagella, is a mouth, or maybe more accurately a portal, where *Trichonympha* draws in wood chips for digestion. That mouth, much like yours, is covered by little jujube-shaped bacteria—a nano-environment within a microenvironment. But you would have no time to think of these wondrous worlds within worlds because the *Trichonympha*'s great swirls would swirl you in, ever closer to that portal, where you would finally be ripped molecule from molecule in this gut within the gut.

Trichonympha owes its name to Joseph Leidy. When he saw the swirling flagella under his microscope in 1877, he was reminded of a show he'd recently seen with ladies partly dressed in fringe. "The arrangement of the long cilia, clothing the body, reminded [me] of the nymphs in a recent spectacular drama, in which they appeared with their nakedness barely concealed by long cords suspended from the shoulders." While we still call *Trichonympha* a fringy little nymph, DNA probes and scanning electron microscopes allow us to see it in a totally new way: some of that "fringe" is actually made of other symbiotic creatures.

FOR MOST OF the history of microbiology, the vast majority of microbes have been untested and unknown because fewer than 1 percent of them can be grown alone in a petri dish. That began to change early in Phil's career. In the 1980s, microbiologists started to pull DNA from whole communities—a clump of pond scum, or a drop of sewage—at once. Then they used computers to sort out bits of genetic code to identify these unknown microorganisms. Phil got his PhD and did postdocs in this new field, studying microorganisms, their DNA, and bioinformatics—a new field of computer analysis. In 2004 he was part of Jill Banfield's lab at the University of California, Berkeley, which took the entire community of microbes living in a pink biofilm on top of highly acidic water that had drained from a mine. The team sequenced all of the microbes' genes at the same time, using computer programs to sort out which genes did what. For the first time, they got a picture of which microbes were in the soup and also what their genes were collectively capable of. They called this process "metagenomics." By 2007, the startling ability of metagenomics to reveal whole environments through their genes was being compared to the invention of the microscope.

I FIRST MET Phil that year, when he brought along Falk Warnecke, a German postdoc, to discuss the metagenomics research they'd done on the paper about Costa Rican termite guts. Falk, who was the first author on the paper, was slight and blond, with pressed jeans and hair combed from a middle part. Where Phil was expansive, Falk was wary and self-contained. Both men rode their bikes to meet me at a coffee shop in Berkeley.

"A disgusting mess of a dataset," Phil said when I asked about the termite paper. "It's like if you were trying to learn about a house and someone gave you the blueprints all ripped up." What they called data was full of all sorts of probabilities and red herrings, overlaps, and errors: Ninety percent of the microbes were found nowhere else on Earth. Half of the genes in the gut were unknown. And, Phil wagered, a lot of what they thought they knew was probably wrong. "Any single one of those forty thousand unknown genes could be a whole PhD for someone," Falk added. "One of the disadvantages of learning so much is you don't know what it means."

Finding meaning in this huge pile of noise has steered their science inward, into a spacey guided meditation. As much as he relied on sophisticated analysis, Phil was at heart a new thing: an intuitive statisticalist. "A few years ago we discovered things with technology," he said, "now it's with imagination." Phil had an almost psychedelic ability to mix reason and data into a kind of floaty improvisational science he called "playing jazz."

Phil and Falk started to talk fondly about the termites they'd sampled. Phil observed that even though we assume the termite is in charge of the guts, it's completely possible that the guts are in charge of the termite. Perhaps, he added, the termite is just a delivery vehicle for the contents of the guts! I loved this idea and egged him on.

Phil upped the ante, angling out to what I came to think of as his metacommentary pose, his eyes darting comically to his coffee. Maybe, he said, our gut microbes are in charge of us—demanding caffeine, say, or salt—fooling us into thinking we have free will and would like a cup of coffee.

He presented this as a joke, but it really wasn't. Phil's interest in the capabilities of microbial communities has inverted his world so that the microbes are real and we seem to be walking husks, zombies captured by our tiniest parts.* We no longer know who's in charge.

Like Scott, Phil was using the termite to chase a much larger evolutionary question, even though he spent his days fretting about biofuels. Termites, he said, are a Rosetta stone for how the world is organized. "It's a neat little system," he enthused. "You've got all of these symbiotic microbes evolving with the termite hosts. It's a simple enough system, but there's an amazing complexity of hosts and dietary habits. You can pose an amazing number of experiments to look at the evolution of these populations in this closed system. Basically the dynamics of microbial communities can move very quickly—antibiotics could wipe them out in an afternoon—but they also stay evolving with the host for millions of years. The dynamics of change in the genome are not constant—some parts change quickly and some stay constant. We haven't gotten a handle on evolutionary speeds." He had another way to put the question: Did the termites get these microbes from eat-

---

*Consider pandas: They live entirely on bamboo, but why? Recent genetic work has revealed that they have the genes for eating meat, but the microbes in their guts prefer bamboo. And so bamboo it is for pandas' breakfast, lunch, and dinner.

ing dinosaur poo and coevolve with their passengers over the epochs? Or did they pick new microbes up whenever they ate a new food?

IN RENO WE met up with Rudi Scheffrahn, the termite expert from the last safari, and his two entomologist friends, rugged men who walked with John Wayne-ish swaggers, one carrying a machete, the other an ax. They worked for a famous exterminating company, and creating an alcohol-infused catalog of termite species in the Americas had been their life's work. In the field, they'd find soldiers, who have more species-identifying features than worker termites, and then, noting the jags on the termites' mandibles or the shape of their heads, they'd do field identifications, drop the specimen in a small bottle of ethanol, tuck it in their waist pouches, and move on.

Rudi himself has a termite encyclopedia in his lab, with thirty-five thousand vials containing termite samples. For research purposes, he also maintained a whole termite nest in a fiberglass shower stall. Life as termite experts seemed exciting: Rudi's friends were once called in to literally de-bug an embassy in Latin America.

Rudi started the trip immediately, driving his friends in a white car and navigating with that same survey from 1934 that he'd used the previous time.* We followed in a red SUV, with me driving again. Our life together quickly took shape: When Anna and Phil started bickering about whether or not Phil was spacey, I played the role of the mother and told them to settle down. Shaomei didn't say much but later she showed me that she'd organized the four of us on a matrix with creative on one axis and excitable on the other. This odd symbiosis and the deep desire to organize everyone and everything into categories was what I loved about the lab.

---

*That survey had been put together by Sol Felty Light, a professor at UC Berkeley, who was equally famous for identifying sea creatures as for identifying termites. His descriptions of soldier termites are elaborate measurements of the width of the head, the length of the head, the length of the head with mandibles, etc. His head shapes are meticulous, with ridges, hairs, and slopes. He loved taxonomy. When one of his grad students mentioned that he'd written a poem about the sea spiders he was studying, Light replied, "You must be very depressed."

We stopped at nowhere places: an abandoned mattress and a dead sheep. Another time a scraggly dry wash with the remains of a dishwasher, a washing machine, a jeep, and an armchair with termite marks on the legs.

While humans had forgotten these places, from the termites' perspective they were Candyland: a world of sugars hidden in cellulose and lignin. To get to these sugars, *Gnathamitermes perplexus*, long-jawed desert termites, wrap the grass blades in mud so they can do their work in the shade. To me the grass plants appeared to be wearing multi-headed turtleneck sweaters. In 1974 an ambitious bunch of researchers measured these mud tubes and found that *Gnathamitermes* were moving, sorting, aerating, and adding nutrients to more than 512 pounds of dirt per acre every year, enriching it with organic carbon, two different kinds of fixed nitrogen, potassium, calcium, and magnesium—all processed in millions of termite stomachs.

It's much harder to measure how much methane termites produce, though researchers have tried covering nests with plastic hats and Teflon bags. An estimate in the 1980s suggested that termite farts were an enormous source of the world's greenhouse gases, but recent work suggests they account for 3–4 percent of methane emissions. Because the soil around termite nests often contains microbes that eat methane, even that amount may be high.

Phil headed off into the scrub, where he'd crouch down like a mantis staring into a clump of grass or debris, wearing a round-brimmed hat smashed way down across his forehead with the chin strap tightly fastened. When one of the entomologists found a colony, Shaomei, Anna, and I would squat down on the ground and suck them up with our aspirators—the little homemade contraptions of tubing and filters Rudi had given us. Shaomei and Anna could sip up the termites and still carry on a discussion. I tended to suck too hard and clog my tube, or not suck hard enough and end up mashing the termite with my tube. While the two women conversed, I mostly managed my spit and listened.

Shaomei said she saw termites as a black box. She had come to the

United States from China to study the microbiology of wastewater treatment plants—which were also black boxes, but large ones. The idea that a pool of sewage was a black box surprised me. She explained that when we treat sewage in holding tanks, microbial processes break it down and eventually clean the water. Sewage goes in and cleaner water comes out as long as some requirements like oxygen, pH, and retention time are met. But, crucially, we don't really understand exactly how or why these giant sewage microbe metabolisms work and why they sometimes crash. Likewise, the termite's gut is a black box for which we increasingly know the parts, and the results, but we don't know exactly how they work.

On evenings in the field, we'd hole up in a motel, remove the termites from their sampling bags, clean them up in petri dishes, gather them into small plastic vials, and drop those into a thermos filled with dry ice. Freezing them fast preserves not only DNA—the stable strings of genetic material—but also the unstable RNA, which can reveal what genes were actually in play at the moment of death. Perhaps if we knew what termites were actually doing in their guts, rather than what they were capable of, we could understand the black box. Sorting and freezing termites is fundamentally tedious and unromantic work, but Phil's team enjoyed turning it into a game.

Where I saw a termite, they saw a forest of elegant phylogenetic trees with chunks of code on a branching evolutionary time continuum. These trees have become everything—not simply a tool for doing work, but a whole way of understanding the world. One day while I was driving, Phil mentioned that phylogenetic trees can be used to sort all kinds of data, not just genes. "Music," he said, "you can clearly see how it evolved if you sort it out on a phylogenetic tree."

In any case, they don't quite see the termite as a termite anymore. Living in a world of computers, code, and symbols, they struggled to connect the termite they knew—a bucket of genetic code—with the eyeless insects running around the petri dish. They settled on imagining them as something like Japanese anime—all that was scary and *Other* became cute and relatable.

While she herded the bugs around their petri dishes, Anna made kissy noises and said, "Here, termey termey." The termites dashed around the dish in a herd, and often after an hour of sorting, and the youthful ebullience of the geneticists, I'd start seeing the termites as six-legged cartoons, too. During one evening's termite-freezing exercise, Anna made a joke about reading the termites their rights before plunging them into the dry ice. Shaomei and Phil giggled. So did I.

That didn't sit right with the entomologists, who'd been drinking beers while we sorted. "Rights?!" said one of the termite terminators. "They don't *have* rights. They're insects!"

The party lapsed into embarrassed silence. It was a joke, but probably none of the geneticists could really imagine a living being without rights. On a previous trip Phil had even asked a farm couple to sign away the intellectual property rights to termites he and Falk had found in a cow pie. The geneticists' world, constructed as it is around databases of genes, is relativistic, not absolute. A living thing is a living thing because it has genes, which makes it different from a rock, which obviously does not have rights. And if you were to put the question to them, they would probably start to muse philosophically about viruses, which have DNA or RNA but are not alive. . . .

Still, the terminator certainly had a point: he was paid to kill termites for the good of society—because there are some living things we all want dead. And perhaps he was using the termite as a delivery vehicle to vent a perfectly valid complaint: biology was leaving the realness of the field to become a bunch of datasets managed by people who didn't know a damn thing about bugs.

There was an awkward moment. During the long lull in the conversation we could hear the plink of the little plastic test tubes—many millions of genes at a go—pinging off the dry ice in the thermos as the things we still called termites met their end.

HANGING OUT WITH Phil and his lab changed me in ways that were subtle and profound. First of all, Phil didn't just study symbiosis; he lived it. Thirty-nine names on the paper; Falk at the interview. Anna always by his side. Whoever the lone scientist is anymore, she is not doing metagenomics. I bought a copy of *Bioinformatics for Dummies*, resigned myself to understanding a fraction of what I heard, and joined in, grateful to be able to watch them work.

And I learned the number one rule of symbiosis: in gaining the benefits of the group, you will also lose things. I had to give up my reflexive pessimism. Phil is deeply, truly optimistic, and he has a conviction— based on the dramatic success of metagenomics—that the world has rules and they matter. My time writing about oil and climate change had torn my feelings and my convictions apart. When people asked me if I had hope, I had taken to saying that hope was not a feeling, it was a strategy. What I meant was that if we wanted improvements, we needed to have hope to push for them, but I personally felt emotionally and rationally pessimistic. Looking back, I think maybe I was speaking in code, even to myself, and I didn't want to acknowledge how hopeless I felt. Unfortunately, nearly everyone interpreted "hope is a strategy" as an upbeat exhortation, which made me feel worse. But for Phil, science itself was a strategy that was hopeful.

At the end of the trip, I drove the SUV down the Las Vegas strip near midnight. Colored lights slid off the car, drunk people tripped across the crosswalk in front of us, and I felt gloomy. Phil was leaning forward intently in the passenger seat, craning his head to look down and then up at the buildings. He was watching the people and the lights with such unguarded delight that he even called his wife to exclaim, waking her up. I said something predictably sour about this being a waste of electricity to get people to give up their money. No, Phil argued. People came here to the desert and built this out of nothing. It's a monument to what humans can do if they feel like it. Las Vegas was a proxy for SCIENCE, that combination of hope and problem solving and luck and money. It was nice to look at the lights from this angle, and I felt much more generous than usual as we watched the fountains at the Bellagio burp flirtily while the sound system played "Hey, Big Spender!"

TEN

LIFE IN THE FIREHOSE

CALIFORNIA

THE JOINT GENOME INSTITUTE, where Phil's team worked, is a low, innocuous brick building in Walnut Creek, California. When I first entered its dark, drab lobby, in 2008, its main decoration was a red digital sign that read: "1.89 B, PASS RATE SHRED Q20 BASS 25, 555, 288 MONTH TO DATE: 1,503,411,395." This was a triumphant, anxious announcement that only geneticists could understand: every second another thousand base pairs—the twinned DNA nucleotides that are the building blocks of genes—spat out of JGI's sequencers. The lab was proud of this—every hour they logged as many base pairs as they did in all of 1998. By the next time I visited the lab, in the fall of 2009, their output had risen eightfold.

Worldwide, the discovery of genes has outstripped the discovery of stars. Every seven months the genomic data on hard drives doubles. By 2015, JGI's output was 1,140 times 2008's, and the cost of reading a genome had fallen from $10 million to about $1,000. At the lab, they called this phenomenal growth in data "the firehose."

Termites were part of the firehose: the lab had servers of data on their genes, but no one knew what it meant. The only way to verify this futuristic genetic data was to go to the "bench," which was what they called workspace in the labs, and try to match it up with the reality of the termites' guts. Phil had a slogan that was only partly a joke: "Sequence First and Ask Questions Later."

With the grant from EBI, members of the lab had thrown themselves into chasing DNA, RNA, and proteins through the termite. Briefly, Phil hunted for those intertwined spirochetes he'd seen in the microscopic photos at the airport, but he didn't find anything useful. One researcher sat in a dark room trying to sequence single bacteria for months, while another tried to reach the same end by washing millions of bacteria across plates covered with tiny micropores. In the background, they explored Phil's Rosetta stone question of how termites got their microbes, and why they stayed.

It was clear that the termite was no longer in the running to provide genes for grassoline—the bug was just too complex—but it had become a sort of mascot, biological proof that those cellulose sugar chains could, in fact, be cracked. The termite could go to the moon, but scientists couldn't yet. For the biofuel project, the lab had turned its attention to wood-eating microbes in compost and in shipworms. But the termite remained a big shining example, an inspiration, and so Phil's team continued to comb termite guts in search of ideas, microbial strategies, and systems.

And the only way to get inside the termite was painstaking benchwork. Falk showed me how to dissect a termite frozen at −80°C. He grabbed the head with forceps and the anus with a pin and, exhaling slowly, pulled his hands apart so that the termite's exoskeleton slid off and the gut uncoiled into a sticky translucent string slightly longer than the original termite. The lump in the string was P3, the third stomach, and it contained a glurp of microbial paste half the size of a sesame seed—a quantity described as a microliter, which is a millionth of a liter or one-fiftieth of a waterdrop—home to millions of individual organisms. To get enough of this paste to sequence, he needed to dissect at least forty-nine more termites.

Benchwork is highly idiosyncratic and archaic, not to mention repetitive, boring, and risky. While Falk was so particular when he worked that he seemed to stop breathing, Anna approached the bench with ferocious confidence, wielding her nine-headed pipette over nearly a hundred vials with a speedy rhythm. She didn't dally: one of her first jobs had been at the USDA, collecting norovirus from human feces. Benchwork has a guild-like aspect: young lab workers are trained to be practitioners in the arts by slightly older workers. Ancient anachronistic tools are used: among the stains used to identify proteins are Coomassie blue and Congo red, their names dating from Europe's conquest of Africa in the late 1800s. Researchers record their results in heavy bound lab books with a design that hasn't changed since before Rosalind Franklin met Watson and Crick. Negative or confusing results are often discarded.

Sometimes Anna didn't bother to dissect termites at all, but smashed them up whole, pronounced it a "nice paste," and spun out their exoskeletons with a centrifuge. She was a daredevil Steve McQueen in white coat and track shoes, flipping her thick ponytail triumphantly. If things failed, she'd have already moved on to a new experiment.

Phil never worked at the bench; he said he anticipated mistakes and got paralyzed by the thought of making one. He told me that he hired Anna when he saw her fearlessness at the bench: they were a good fit.

Knowing the risks, I found watching benchwork to be paradoxically relaxing, like watching Olympic-level ice dancers. Even though it made up a tremendous amount of the labor in the lab, no one really doubted that this work would soon be done by robots; commercial biotech labs already had machines that could run millions of samples at a time.

When the lab sequenced the guts of the *Nasutitermes* who lived in the shower stall in Rudi's lab, dining on cellulose sponges and pepperwood, they found evidence of genes for some large proteins that might be involved in breaking down cellulose. What were these proteins and what did they do? One day in the spring of 2010, Shaomei asked if I'd like to join her in the hunt for them.

WE MET A few days later at the Joint BioEnergy Institute, the government-funded biofuel research center in Emeryville, California. Shaomei was grinning, dressed in fleece as though going out for a run, and carrying frozen termites on dry ice in a Trader Joe's bag. We went into JBEI's lobby with its high ceilings, bas-reliefs of Nobel Prize winners, and a large Buddha. No anxious readout here: JBEI has taken its cue from commercial biotech to project expansive optimism. On the upper floors we walked down hallways with shimmering bamboo floors and windows treated with bright green film the color of cellophane grass in Easter baskets. JBEI feels like a charmed world.

A molecular biologist named John was waiting for us in the proteomics lab. John had a shaven head, a Chinese-character tattoo, and his workbench was the usual stark black, with racks of vials with hot-pink and orange tops, orange biohazard stickers, pale-pink and green marking tape, and yellow and purple stoppers. A bank of fridges hummed off to one side. One was labeled prominently "No Food," which struck me as funny. There was a whiff of something living, a breath of some sort, around the lab. We put on biosafety glasses, which gave our heads square translucent corners like Minecraft characters, and set about gutting eighty termites.

Shaomei was faster at gutting than me. What I found difficult was the discipline of keeping my mind to the task at hand. It's a forceful meditation so that the pointy ends of the forceps become my fingers, and my eyes are tuned to the most minute distinctions in the pale termite. Let go of this focus and I start to swell so that I'm rising from the ground to hover over the dissected termite like a Mickey Mouse balloon in the Macy's Thanksgiving Day Parade, my fingers bulbous, stiff, and suddenly useless. If I allow myself to fill the room, I won't be able to stuff myself back into the vial that dissects termites. I mention this because the need to control one's head is a regular part of life at JBEI.

IN 2005 RESEARCHERS at the Department of Energy had estimated that if the United States went totally termite we could harvest trees, crop residue (such as cornstalks), and high-energy grasses, and engineer microbes to turn them into sugars. Then those sugars could be fermented to make nearly 60 billion gallons of ethanol—a potential gasoline substitute—a year by 2030. In 2016 that estimate was updated to 100 billion gallons. Theoretically—and all of this was very theoretical—that would equal most of the petroleum we used for driving in 2015, while reducing greenhouse gas emissions from driving by as much as 86 percent. At the time, it was not possible to process cellulose into sugars in great quantities, and what was possible was much too expensive to burn in a car.

JBEI's explicit goal was to brew biofuel at a price that could eventually compete with gasoline. To accomplish that, the lab needed to engineer biological processes so that they are predictable and can scale from the small flasks in lab experiments to vast industrial tank farms. Teams of researchers focused on understanding and manipulating the plants themselves, understanding and increasing the processes that can break down cellulose, and designing microbes that can synthesize fuels from the sugars. JBEI's CEO was Jay Keasling, a leader in a new field called synthetic biology that is working to engineer biological systems—cells, enzymes, metabolic pathways—so that they produce valuable molecules.

The future of biology Keasling imagines is modular and LEGO-like: "One can even envision a day when cell manufacturing is done by different companies, each specializing in certain aspects of the synthesis—one company constructs the chromosome, one company builds the membrane and cell wall (the 'bag'), one company fills the bag with the basic molecules needed to boot up the cell." The implications of this transformation will be huge, but at the moment the work is at the bench, minding one's head.

HALF AN HOUR LATER we had a gob of goop in a chilled vial. John gave Shaomei a sixteen-step "recipe" to break open the bacterial cells, get rid of salts and fats, and separate out the proteins using methanol and chloroform and a centrifuge. After two hours she had a vial with two layers of liquid separated by a white disc. The top layer of liquid was water, which presumably had the proteins dissolved in it, and the white disc was salts. She carefully pipetted off the water into a new tube, added methanol, spun again, and got a small dusting of brownish stuff that John said was probably proteins. She dissolved them in a purplish buffer solution that contained a detergent with a negative charge to bond to the proteins. Then she heated the tube, and the combination of heat and detergent unkinked the proteins so they lay in flattish chains rather than in their characteristic 3-D structures.

Shaomei went to one of the big fridges to get a gel, which is properly called an electrophoresis gel. It is a piece of clear gelatin a little bigger than a deck of cards that works as a sophisticated sieve that can be used to sort molecules of different sizes. Shaomei injected the purple protein solution in lanes across the bottom of the gel. When she was done she injected a test marker full of mixed-weight proteins to confirm that the gel was correctly loaded. The dye clustered in the end of the gel where it was injected—it would need an electrical current to move. She loaded the gel into an acrylic box that had a red wire coming from one end and a black one from the other. John helped plug in these wires and turned the current to 50 volts and then to 120. Because the proteins were negatively charged by the detergent, the current pulled the proteins through the gel, with the smallest proteins moving quickly while the larger ones moved slowly. After electrophoresis, the gel was incubated in a Coomassie dye solution to stain the proteins and then it was washed. In the end, purplish bands of proteins were arranged in a sort of ladder from largest to smallest. The whole contraption was called XCell SureLock—a name clearly chosen to instill confidence in postdocs.

As soon as the electricity was turned on, Shaomei sat down to record everything in her black-bound lab book. There is a mismatch between the precise and numbing repetition in the lab, the big questions they're trying to address, and the great likelihood of being left with nothing of value. I understand why Phil can't stand the anxiety of the bench.

But postdocs have no choice. John said that one year in graduate school he counted worms without talking to anyone all day long. He just looked in a microscope and hit a counter. His year of work turned into a single line in a large paper. But he was lucky, because researchers only get credit for these heroic feats if they got a positive result. Shaomei said she got nausea and vertigo from looking in a microscope for hours. John said, "I'd go home and count worms in my dreams."

Synthetic biology might someday move us to a new chemistry where microbes in vats produce the molecules that we now extract from coal and petroleum, but for young people in both chemistry and biology it's causing angst. "There's a breakdown happening, a disciplinary crisis about what our 'knowledge' is," a chemical engineering student at UC Berkeley explained to me earlier. "The chemistry we grew up on was distillation columns, but now it's going to be Keasling and synthetic biology." But, like the roboticists, the biologists and chemists alike were "Waiting for Carnot," and trying to keep their minds devoted entirely to their small experiments while the greater dream of a different system took shape under these weird green windows.

The next morning the lab had an unhealthy, warm, seething smell that John attributed to the compost-derived microbes living in his petri dishes. In my lifetime, chemicals have been recognizable by their petroleum scent, but as chemistry reorients toward biological methods, that smell will go away and be replaced by these strange live smells that seem to communicate with us in primeval, subconscious ways.

Over on the bench, the gel was successful—little purplish stripes appeared in the lanes. By comparing these stripes, Shaomei could estimate the molecular weight of the proteins in the gels. She cut each lane into twenty-five squares and set about extracting the proteins from each tiny square. It was another day of laborious processes on increasingly minuscule quantities of gray crumbs.

When it was finally extracted, the protein—it was just a squidge of stuff now, barely visible—was sent off to the crew who worked with mass spectrometry. They would hit the proteins with an electron beam to determine the identity of the amino acids and then use that to make educated guesses about the likely shape and identity of the protein. The thought of this made John philosophical. "We really don't understand how proteins work. We know that they're made of amino acids but we don't understand how they fold. They have a pocket here and a pocket there." A protein may behave one way in acid and another in water. His frustration is tempered by the implications of the firehose: recently a nearby company had identified a chaperone enzyme that speeds up other enzymes.

What we didn't realize that day in the lab was that another big discovery was hiding in plain sight. In 2007 Phil and two members of the JGI lab had written a paper on a strange repeating series of nucleotide bases that seemed to protect bacteria from viruses. These "clustered regularly interspaced short palindromic repeats," aka CRISPRs, had been first found in E. coli in 1987. The lab's paper looked at how these palindromic genetic structures were found in 195 different bacteria, suggesting that they were present in 40 percent of all bacteria. But in 2009, they were just a funny feature of bacteria. Phil and the other researchers moved on to new things.

In 2012, researchers at Berkeley, Harvard, and MIT realized that these palindromic features could be coupled with an enzyme called Cas9 to edit the genes in cells, including some human ones. This discovery gave rise to a patent war and tens of thousands of papers over the next few years, looking at everything from how to edit human egg cells, to how to insert "gene drives" that would reproduce within cells to spread specific DNA through whole populations of malaria-carrying mosquitos, to how to protect your breakfast banana from a deadly fungus. CRISPR/Cas9 seemed like a tool that would enable humans to change virtually everything, a way to genetically engineer the living world in the way we wanted, but it had been sitting there all along, just a funny palindrome in the firehose.

I PLANNED TO go back to the lab with Shaomei, but on April 20, oil and gas exploded out of BP's Macondo well in the Gulf of Mexico. Spreading across hundreds of square miles of Gulf waters, the spill made visible the ugly reality of our complicated dependence on oil. The fishing and oil industries in the Gulf were shut down. Ten days later oil was washing ashore in Louisiana. The spill went on all summer, with the ruined underwater wellhead spewing on TV around-the-clock. The disaster easily made the argument that we should invest in biofuels: these neat low-carbon substitutes could be swapped into our transportation system without drastically changing the cars or our political conversations.

Synthetic biology has somehow skirted the controversy that has dogged genetically modified organisms—perhaps because modified slurries in tanks are perceived differently than modified potato plants growing in a field, but most likely because few people have heard of it. But many of the scientists working on the field see themselves as trying to fix the world. Of these, Jay Keasling, who is soft-spoken and grew up on a farm in Nebraska, is a leading light.

In 2002 or so, Keasling learned that one of the few effective malaria drugs, called artemisinin, was rare and expensive because of shortages of its main ingredient, sweet wormwood. He reengineered the bacterium *E. coli* to make a precursor chemical for artemisinin to replace wormwood. He hoped that this process, reengineering cells to produce this complicated and costly molecule cheaply, would provide cheap drugs for the 300 million to 500 million people a year who get malaria. To that end, he got involved in a huge philanthropic project, with more research funded by the Bill and Melinda Gates Foundation and a deal with the pharmaceutical manufacturer Sanofi to forgo profits when distributing the drug. This seemed like technology above reproach: solving the world's problems.

The anthropologist Gaymon Bennett spent time embedded at a project within JBEI, watching scientists work. He wrote that synthetic biology is based on the provocative idea of "responsible innovation," which is a term of art for social scientists. What struck me, though, was his description of the way the discipline "relentlessly imagines human problems as uniquely susceptible to biotechnical intervention, and imagines itself as uniquely capable of delivering that intervention." Artemisinin appeared to be the perfect example of the premise of responsible synthetic biology.

But this wasn't the only vision for the field. A month after the oil spill, Craig Venter announced that his lab had created a cell named Synthia with "chemically synthesized" DNA. Mixing the language of data and biology, he said: "This is the first self-replicating species we've had on the planet whose parent is a computer."

Venter, the flamboyant founder of several companies and a nonprofit institute, does not come across as humble or introspective like Jay Keasling. In 1998 he challenged the government program to sequence the human genome to a race, which led all parties to speed up dramatically, producing JGI's anxious red readout of their daily progress and the firehose itself.

Venter sees synthetic biology as a way to leverage the scaling ability of microbes to let us live without the limits of traditional resources. In 2011 I saw him speak in Washington, D.C., and according to my notes, he said: "My view is that we can replace a whole system of manufacturing with biology. With biology, increases of five to tenfold are common. Even an increase of a millionfold is possible." At its heart, his argument for synthetic biology is the same as the one against it: the old horror movie line, "IT'S ALIVE!"

Venter's speech was a mix of rhetorical strategies, one minute summoning up the magic of life, the next promising to rationalize it until it's a machine: we'll "digitize biology," put code in cells and "boot it up." These comfortable laptop-like metaphors hide the scale of the transformation he's talking about. "If we can make food a thousand times more efficiently we'll design food in new ways, in a far more nutritious fashion." That made him think of another idea: "Why not get rid of the cows?" Synthetic biology, in Venter's vision, is a fountain of limitless abundance: tanks of microbes growing endless hamburger.

To his credit, moving from cows to tanks has the potential to dramatically reduce greenhouse gas emissions.

In the summer of 2010, President Barack Obama convened a Presidential Commission to discuss the ethics of Venter's cell, Synthia. The commission worried about making a decision that would hold the field back from its potential. Months later, they made eighteen recommendations, mainly that scientists and labs should police themselves with "prudent vigilance" and biologists should take ethics classes. And so the ideal of "responsible innovation" was a substitute for regulating a science that hadn't arrived yet.

SHAOMEI AND JOHN'S three days in the lab amounted to nothing. The mass spectrometry team found some proteins in the sample, but they didn't line up with the proteins that had been found before and had predictable functions. The databases simply weren't large enough to cover all the genetic potential of the bacteria in termite guts. The results of the work were not recorded.

ELEVEN

JAZZ IN THE METAGENOME

I N MAY 2010, PHIL INVITED ME out to JGI to watch the team do a session of playing jazz, which is what he called the process of interpreting the data from the termite guts. The team met in a bare room with the lights off and the shades drawn. The three long afternoons I spent with them may be the closest I'll ever get to the feeling of actually *being* in a termite's gut. It was more than the fact that both the room and the termite's gut are mostly airless: members of the lab—Shaomei, Phil, the senior researcher Natalia Ivanova, and the postdocs Martin Allgaier and Amrita Pati—lost their distinct identities and merged into one being with multiple hands and brains. Beyond a mind meld among the scientists, it was also a free-floating jam session between humans and multiple computer databases, crossing a technical boundary that I wouldn't have thought possible before I entered the room.

Martin began projecting a database of termite gut microbe genes from the metagenomic surveys on a wall. Along the right-hand side was a list of different likely genes. Reflecting the uncertainty of the data, which was based on statistical probability and guilt by association, the genes were referred to as COGS or Clusters of Orthologous Groups. The genes had names that were a series of letters and numbers, and each gene provided the code for a protein. The next column explained that protein's assumed functions (breaking down cellulose, for example, or the ability to fix nitrogen in a way that's biologically useful).

The group was going to compare the metagenome of the wood-eating termites that lived in Rudi's shower stall with the termites we'd found living in cow dung in Arizona in 2008. At any one time, we could view only a dozen out of 470,000 protein-coding genes in the two termites combined, so Martin kept scrolling down the page while we leaned back and tried to absorb this information intuitively. The process of finding some sort of gestalt amid this pile of interesting but scattered information was difficult. But the real issue was what the scientists couldn't see: Of the protein-coding genes in the dataset, fewer than 40 percent had a predicted function. So we were missing 60 percent of the information.

When the team happened to come across an interesting gene and wanted to see where it occurred in the wild, Shaomei would fire up a ghastly UNIX shell program with a navy-blue background that displayed genes and their owners in primary-color text, arranged in barely readable phylogenetic trees explaining how similar genes were found in, say, bacteria, yeasts, flies, and humans, followed by letters of inscrutable internal codes. Phil looked at the lengths of some branches of the phylogenetic trees and declared the data "shonkey." This was the firehose in action: as JGI's collection of microbe genomes quintupled between 2009 and 2011, some "shonkey" data made its way into the database along with the good stuff.

This exercise hit me as a shock. My experience of metagenomics so far had been reading published journal articles, which seemed to provide hard evidence—even a godlike perspective—for conclusions I had assumed were grounded in data, if not fact. Yet the database screens were inscrutable and the information seemed compromised from the start. Frequently, we'd get to the third or fourth level of detail and Phil would look closely at the sequence and say that it didn't make any sense; it was miscategorized, the product of some earlier error. He worried about "genome rot"—when bad guesses piled on top of bad data, and eventually a whole tree of suppositions could collapse in on itself. In order to do this work, the scientists had to be comfortable with knowing very little for certain. I came to think that this was one of the biggest differences between being a scientist—this comfort with multiple unknowns—and being the rest of us, who rely on narratives that suggest we know much more than we do.

I had to learn to enjoy not knowing because I often had no idea what the researchers were talking about. They spoke a specialized vocabulary of Pfams, BLASTs, and other acronyms that I didn't understand in a spectacular range of accents. I frequently gave up on comprehension and just surfed the rhythm. But I stuck with my camera and my notebook, and a story did emerge.

The first obvious thing about the two termites was that they had very different organisms in their guts: the dung eaters had lots of bacteria called firmicutes (which make up a lot of the population of human guts), while the wood eaters had 70 percent spirochetes, with their wiggly tails. What was really interesting was that despite the difference in the identities of the bacteria they held, both termites' guts contained genes that coded for strikingly similar functions.

This simple observation is revolutionary. Previously, scientists tended to think of genes as the property of specific organisms, but metagenomics allows us to look at communities and even whole ecosystems as essentially pools of genes and possibilities. The metagenomic view shows that termites have guts that do certain jobs—think of it as a spec sheet for eating wood: soften the cellulose, chop up the sugar chains, ferment the sugars, and so on. All of the microbe species who've evolved for the party in the termite's gut end up playing along with this essential script. And in doing that, they lose genes that they'd have needed to survive independently outside the gut and gain genes that allow them to be more helpful inside the gut. Finally, they are capable only of living in this one termite gut environment.

We can see this right on the screens, even though we never saw a photo of this spirochete, or any evidence beyond its likely genes and their place on phylogenetic trees. Phil got the group to flip between databases to get the genomic data from a single spirochete, which strangely lacked its usual kit of genes for mobility and tracking toward chemicals. "What's going on? This is totally atypical for a spirochete!" said Phil.

Moving and sniffing for chemicals are defining characteristics of spirochetes. What is a spirochete that can't move or smell? It's an absurdity, and yet it is right there, in the data. Shaomei wondered if the spirochete's genome got smaller and lost its genes for defense and mobility as the spirochete spent more evolutionary time in the termite's gut. Phil hunched inward in front of his computer and then looked up to announce that this particular spirochete is living inside a protist—like *Trichonympha*—which lived inside the termite.* Protected inside two different organisms, apparently it no longer needs to move or defend, and so has lost those genes. Once you go symbiotic, you can never go back.

It's here, in this stuffy room, that I can see for the first time what it means that the termite's gut is another composite animal made of millions of bacteria, who, like their termite hosts, have traded away eyes and wings for the advantages of living in numbers.

Outside this room, metagenomics is driving a lively new debate on the old question of the role of altruism in evolution. This question, which goes back to Eugène Marais's musings on the superorganism and E. O. Wilson's work on why insect societies stick together, comes up again at the microbial level. While competition has been part of the evolutionary process, at the microbial level it increasingly appears that

---

*In theory, protists should not have been living in the *Amitermes*, which is a higher termite, but they have been found there. C'est la vie termite.

cells compete to cooperate in communities—fitting in and helping out is essential to their survival. The microbial philosophers Maureen O'Malley and John Dupré wrote that while Darwin's general picture of the influence of competition on evolution is correct, it appears that "contrary to the orthodox evolutionary view that altruism is exceptional and requires special explanation . . . the norm among organisms is a disposition to act for the benefit of other organisms or cells." To get ahead they've got to get along. Codependent forevermore. Our old friend the superorganism has shown up here too, though sometimes it's called a meta-organism.

One conclusion I drew was that termites may abide. The massive redundancy in their genes, the ability to fine-tune their metabolisms to live on the crummiest (or richest) foods, are survival strategies for any eventuality. We tend to think of evolution providing elegant minimalist solutions to problems, but in the case of the termite's gut (as with the way termites overbuild their mounds) there is massive redundancy and duplication. Gene-wise, a termite's gut is a factory with a thousand extra workers standing on the sidelines.

This is particularly obvious in the case of nitrogen. Nitrogen makes up 78 percent of the atmosphere, but it drifts around in pairs of atoms attached by triple bonds. Living things need nitrogen to make proteins, but we need single atoms. In nature, those are produced by lightning strikes and bacteria with enzymes that do the job of breaking up the twins and pairing them up with hydrogen to make ammonia. Termites' guts generally contain lots of bacterial genes for fixing nitrogen.

The biggest difference between the wood-eating *Nasutitermes* from Rudi's shower stall and the *Amitermes* who lived in an Arizona cow pie was that the wood eaters have tons of genes for fixing nitrogen while the cow-pie eaters don't. This isn't surprising: wood is a nitrogen-poor food, so the wood eaters would need ways to fix it for themselves. Cow dung, on the other hand, is rich in the stuff (because the cow's stomach microbes have already gone to the trouble of fixing the nitrogen). So somehow, termites' food sources may influence the capabilities of their guts. But how?

The nitrogen problem was Phil's Rosetta stone question—how did the termite get its gut creatures, and why do they stay? The only way to figure this out was to begin checking some of the individual nitrogen-handling genes in the gut.

As Martin scrolled through the database of genes, Natalia began defining what they did, speaking out of the side of her mouth, through clenched teeth. Natalia was closer to Phil's age than to that of the post-docs in the room. She was outwardly reserved, with curly auburn hair, but when she smiled, her eyes turned into crescents. Now she was not smiling, and when she spoke her jaw did not move at all. She seemed to be in deep communion with the database. After a few minutes I realized that Natalia had an eidetic memory, and she was taking the gibberish of the protein names, like Q64B54 and naming their metabolic function. Martin and Shaomei did side searches as she continued to narrate the seemingly random code. The people in the room were merging into one. Or, like the microbes, they were simply sharing functions and tasks.

We flew through the codes, assembling stories, testing them, and leaving them behind. We had finally left the constraints of test tubes, pipettes, and the whole messy voodoo of the bench. Now we moved godlike through the codes for genes and proteins, a magic storehouse of knowledge waiting to be discerned and understood.

As they moved from gene to gene, anticipating each other's questions, Natalia and Phil began speaking alternately, as if one monologue was coming out of two mouths. After four hours the room had grown stuffy, and the chairs uncomfortable, but at times the discovery process was so exciting Natalia seemed to stop breathing. I realized I was clenching my own jaw.

Phil searched for connections and Natalia nixed them. Sometimes they talked past each other and at other times they were in perfect sync. One moment Natalia's eyes lit up and she spoke with sudden vehemence, turning her body in her chair, with a big smile, as if sleepwalking or in a fugue state. Phil unwound from his hunched position, arching his head upward. There was something about the whole experience, the intuitive professional elegance of it, that reminded me of Fred Astaire and Ginger Rogers dancing. But whatever was going on actually included four people. Martin's and Shaomei's hands were also part of the dance. The database swooped this way and that like a bass line. This, then, was what Phil meant by "playing jazz."

The process was beautiful, but it was a little haphazard, just like Phil's rueful joke about sequencing first and asking questions later. "It's an unsettling feeling the first time you do it," he told me once. "Essentially you feel you're walking around groping in the dark and out of that dark would emerge . . . *something*. Because these metagenomic datasets are so big and there's so much data, you have to do a lot of cross-referencing with different disparate information, making lots of big leaps." It also takes a certain discipline to push theories and test them again and again. "I reckon the educational system teaches you to think in a rut, so there's a certain activation energy to throw yourself out there and actually say something untested that might push the ideas."

The sprinklers came on outside the building after five, but I was in no hurry to leave. Databases are an unlikely spot for jazz, never mind anything fun, but—like sewage processing plants—they are one of the invisible foundations of our society. If I were going to college again, I would get a degree in microbial ecology, and I would surf and shimmy and jazz my way through piles of data and down half-pipes of possible connections. This molecular narrative—with all of its millions of years of evolutionary possibilities—was infinitely more satisfying than writing. It was real and it was psychedelic. I was a convert.

During these meetings I decided it was absolutely necessary for all well-educated people to know how to read a phylogenetic tree—as core as, say, knowing a little Shakespeare or some economics. Afterward, Phil described the thrill of finding a plausible story about the relationship between the organisms and their genes. "Something starts to form and you think that's too speculative to be real, but then it solidifies. You build enough of a case to write a paper and somebody else finds something and suddenly five or ten papers down the track that association is like concrete, hard stuff, science, a fact!"

At the same time, the process was obviously random and unsystematic. Only the highest-level players—Phil and Natalia—could impose any kind of intuitive order on the data at all. And even then, it was essentially a high dive they'd trained themselves to do. Whenever I thought about what I watched in that room, I thought of Natalia. Could anything have happened without her? Can a science run on the minds of rare Natalias?

And metagenomics itself supplies information that can be misleading without context. The methods used to sample environments may amplify the most common organisms, causing researchers to miss those that are less common, but potentially important. Consider *Crocosphaera*, a single-celled cyanobacterium that lives in seawater, does photosynthesis, and also fixes nitrogen. Researchers initially focused on seawater microbes that were abundant, but when they found fragments of genes from *Crocosphaera*, they realized it was a key player in ocean ecosystems. "It's another leap of faith to find the little ones," Phil explained.

And then the genomes themselves are enigmatic, and do not reveal as much as we think. At Phil's suggestion, I called Patrick Keeling, a microbiologist at the University of British Columbia who's been studying termite guts for more than twenty years. If we only looked at genomes, he said, we wouldn't know that crows can use tools. We might not even realize they can fly! But with microbes, genomes are especially misleading because they don't reveal two important things: behavior and structure. *Trichomonas termopsidis*, for example, processes wood in termites' guts, but in a vagina its close relative *Trichomonas vaginalis* is an STD, eating vaginal secretions. The genomes of the two are similar enough that it would be difficult for scientists to understand how differently they act in the world.

By the end of the third session, the team had an outline for a paper that Shaomei would research and write. It was a tale of two termites with different microbes but similar genes and functions. The way that they diverged was in their food: the wood eater had tools for eating wood, making its own nitrogen the way some people carry their own hot sauce in their purses. By contrast, the poo eater had tools for eating nitrogen-rich food. Still the Rosetta stone question was unanswered: How had the termites gotten their microbes, and why did they stay?

Phil didn't get the answer to that until later, after he'd moved to a new lab at the University of Queensland in Australia, where he assigned an honors undergraduate named Nurdyana Abdul Rahman the task of comparing microbes in the guts of sixty-six different termite species. By this time, new analytic tools enabled her to analyze a huge volume of data. Dyana became fascinated by the project and used it as part of her PhD thesis. Her paper, published in 2015, had just eight authors, compared to the thirty-nine that were listed on the first metagenomics paper in 2008. She found that termite gut microbes coevolved with their termite carriers over time, swapping functions among the different organisms. The termites didn't pick up new organisms; the termite and the gut microbes changed together. When their diets changed, it appeared that the termites could rebalance their gut portfolios without changing the list of inhabitants, only their relative numbers.

So the answer to the Rosetta stone question was that termites and microbes lived in deep symbiosis over millions of years, becoming inseparable. The amazingly wide numbers of genes doing similar things in the gut seemed to allow the partners to adjust to whatever the world threw at them.

While it was interesting to know how the termites and their bugs evolved, it was still an open question whether a system so tightly bound together, so self-regulating, could be disassembled to reliably produce products such as biofuels. The ability to swap genes and change behaviors has been key to the survival of the termites and their symbiotic fellow travelers, but they remain more like superorganisms (with all their cultish connotations) than gene-based computers.

Biofuels hadn't yet given us the powers of termites, but the firehose of microbial information did break out of the lab and invade human social mores in the strangest way. I hesitate to mention this, because it is gross, but I can't resist: Americans started practicing a form of trophallaxis. Following instructions copied off the Internet, thousands of people began performing at-home "fecal transplants." In medical trials fecal transplants had been successful in helping people get rid of stubborn and sometimes deadly *Clostridium difficile* bacterial infections in their guts. Why this procedure worked was unclear and similar to the Rosetta stone question: Why does the "*C. diff*" bacterium take over, and why does it sometimes give up when a new community shows up in the human gut? More-recent work suggests that the success of some transplants has to do with proteins and viruses that live in feces more than the actual bacteria themselves. As word of the treatment spread on the Internet, the number of such transplants being done at home, using kitchen blenders and equipment from the drugstore, grew to be far greater than those done in medical settings. This mixture of science and folk medicine encouraged people to overcome centuries of inhibition as they welcomed neighbors shyly holding a plastic container of fresh, warm "transplant material." I wonder what William Wheeler, with his interest in trophallaxis, would have made of this.

I saw Phil the day before he left for Australia and mentioned that while I enjoyed being in the jazz sessions, they seemed like a tough way to do science. "It's a descriptive discipline and it's very anecdotal," he lamented. "Microbial ecology will be in a golden age when it becomes a predictive discipline." It was far from that now. Even with all the data in the firehose, we couldn't look at the microbes in a person's gut and discern whether the individual was healthy or about to be ill with Crohn's disease.

"You have to talk to Héctor," Phil said. "He's a physicist and he's trying to drag microbial ecology kicking and screaming into a predictive science. He has a slide about a car. He's very passionate and he has a chip on his shoulder."

TWELVE

BURNING VERY SLOWLY

A FTER PHIL LEFT for his new lab in Australia, the idea of the termite as a model for biofuels was pretty much dead, at least at this lab. Still, I wondered how scientists working on biofuels imagined we'd get the capabilities of termites—not to mention un-limited growth and solutions—from clots of microbes in stainless steel tanks. I decided to take Phil's advice and give the physicist Héctor García Martín a call.

We met in June 2010, on a balcony outside of JBEI's optimistic green windows, where he immediately let me know that he thought Phil's "jazz" sessions were ridiculous. When he was part of Phil's team in the early days of the termite research, sitting in on free-association sessions drove him berserk. "We could almost describe how a metabolism worked, but we couldn't predict it! People in biology say, So what? It's unpredictable. I said, my ass it's unpredictable, and I made it my job to piss everyone off by showing them wrong on this account."

Where Phil is expansive and slightly askew, Héctor is compact and perfectly groomed. That day in 2010 his beard followed his jawline almost to a tolerance of a single hair until it reached a perfect spade-like tip in the middle of his chin. He often dresses a little more formally than the other scientists—a shirt and trousers to their jeans and T-shirts. He speaks emphatically in complex paragraphs, delivered in a continuous stream punctuated by shouts of amazement. I never talk to him without my tape recorder. As the lone condensed matter physicist working on the biofuels project at JBEI, he jokes freely about his life as a Don Quixote character, tilting at windmills.

I soon understood why he rubbed the biologists the wrong way; he fundamentally disagreed with the project of observing and describing life. This emphasis, he says, has led to poor definitions of what life is, a clumsy sense of evolution, and an occluded concept of biology's potential. If physics had stopped where biology did, he says, we'd still be watching apples fall and exclaiming over gravity rather than harnessing the principles of physics to design things like cell phones, power plants, cars, and of course bombs. "We've reached the limits of the reductionist approach."

Héctor wanted to de-emphasize the study of individual genes and organisms to look at the big, underlying system enabling life: metabolism. And metabolism to him is a slow combustion. As fire is a violent chemical process, metabolism is life's very low flame. "We're all basically burning very slowly."

When I asked to see what he meant, he showed me a flowchart of how the termite's gut breaks down wood that looked like a map of the Tokyo subway system. Near the center was a loop with hundreds of subsidiary reactions hanging off the sides like intersecting train lines on the Yamanote Line. Among those interconnecting lines were the two different nitrogen cycles Phil and his crew came across during their jazz sessions, but they were just two tiny nodes in a vast network. When I started at the top of the page and traced the reactions downward, I could see the whole process of turning wood into sugars. It would have been even better if I could have understood the chemical notation.

And while the chart made these processes visible, Héctor lamented the lack of laws—a sort of biological thermodynamics or basic set of rules—for how metabolism works. I wondered how many other physicists were hanging around biology labs Waiting for Carnot, as Justin described it. And in the meantime, he was making biofuels. On the day we met, Héctor's position was deputy director of host engineering in the fuels synthesis division of JBEI, and his goal was to make the metabolic map of turning wood into fuel as smooth and predictable as the Tokyo subway system itself.

When I asked him what he thought about termites, he said it would take twenty years to understand them, and for now he needed to work on just a single organism—a nice tame *E. coli*, say, or a yeast. Once scientists had a predictive grasp of the metabolism for a single microbe, then they could think about multi-microbe environments like termite guts. In order to become termites in the industrial sense, in other words, we'd have to abandon the idea of the real bug for a few decades.

HÉCTOR LEFT BILBAO, in the Basque region of Spain, to get his PhD in condensed matter physics at the University of Illinois with the goal of "changing the world." Because condensed matter physics was able to model and predict whole systems, he thought it was likely to produce the next transformative innovation, something like the transistor. He also had a sense that the tools of the discipline might somehow transition to understanding biology in a more holistic way, making it predictable, and therefore useful and world-changing.

Over the course of his PhD he applied physics and math to classic problems from biology and eventually became interested in microbes. In microbial ecology, he saw a way to view ecology and evolution at the same time, and in a rapid timescale. The problem with studying evolution in multicellular organisms is that their genes evolve slowly.

Likewise, to study the ecology of a forest, you need several human life-times to watch it mature, never mind evolve. Microbial ecologies—like those found in termite guts or hot-spring pools—experience both rapid evolution and rapid changes in ecology.

For Héctor, this was a perfect system for figuring out how to model the whole, rather than the parts. He had many ways to analyze these systems. He could look at them metagenomically by all of their genes; he could look at them metabolically by how they acted as a group; he could watch them evolve singly and as a group. Using different types of data, he could zoom in on a single reaction and zoom out to the whole community, playing time backward and forward. Here was a situation where he might be able to figure out how the rules of the whole complex system translated to the parts. What Héctor was after is a variation on Radhika Nagpal's goal of understanding the relationship of global-to-local, but applied to the chemistry of microbes.

I WONDERED IF the scientists I was meeting would get along. Would Scott and Héctor—both against reductionism, but one believing in the striving of the bug and the other in the primacy of the chemical metabolism—get into a fight over their separate ideas of what "biology" is and how to do it? Would the roboticists, with their interest in repeatable data, throw up their hands when they saw Phil's team playing jazz?

Though I started with the assumption that I was watching scientists watch bugs, I came to understand that I had a ringside seat to a much larger, multidisciplinary argument about what life is, and what its relevant units are—genes, individuals, superorganisms, or metabolisms. As scientists from other disciplines enter biology, both to understand and to repurpose it, fights like this are occurring in labs and journals all over the world. When I started watching the researchers at work, I expected that some would turn out to be "right" and others proven wrong—producing a "just so story" about science and rationality. But as the years went on I realized I was actually watching the

great global termite mound of science—a collection of equipment, am-
bitions, ideologies, grudges, blind spots, and insights—interacting and
reshaping the way we think about life.

Héctor joined Phil's lab at JGI as a postdoc, studying termites and
sewage sludge. In the termite meetings, biologists pulled together that
diagram of the gut's metabolism and set about describing which genes
did what. Héctor felt they were missing the point: they couldn't predict
what was really going on. He felt the biologists were looking at a
dynamic system as a series of disconnected frozen moments—literally
that second when the sample took its tumble into the flask full of liquid
nitrogen. These were not little problems, in his view, they were existential: scientists couldn't predict which conditions led to life and
which to death. And if you can't model—or at least explain—the difference between life and death, then what exactly is the discipline
of biology doing? "Physics has evolved to deal with complexity and self-
organizing systems, but biology still treats the subject as a parts list!"

The biologists weren't interested in predictions at all. Héctor was
irritated by his colleagues' stubborn belief in the unpredictability of
biology. "No one was taking what I was saying seriously, so I made a
slide just to piss them off." Aha. This was the slide that Phil had mentioned. On one side was a photo of an Alfa Romeo with facts like its
speed and acceleration. "I told them this is a system based on a hundred years of engineering based on two quantitative predictive
sciences—physics and chemistry." Thermodynamics, the chemistry of
combustion, the behavior of materials under temperatures and
stress . . . all of these understandings go into making a car with a max-
imum speed of 200 miles an hour.

On the other side of the slide he put the biological process used to
remove phosphorus from sewage sludge in thousands of treatment
plants all over the world—what Shaomei had called the black box. But
here, the characteristics—such as the pH and ratios of ingredients—
can change over three orders of magnitude, and the entire system can
crash without warning or reason. "No one knows why it crashes! This
is a system based on thirty years of engineering based on a science

which is somewhat quantitative but not predictive!" Sewage treatment plants, from this vantage point, are a big box of stinky irrationality, waiting to be solved and tamed.

I have to interject here that this is a little unfair. The reason so many of us can live in cities is that sewage sludge regulates its microbial stew itself: it's a superorganism of sorts, with internal feedback mechanisms and ways of evolving to deal with new circumstances over time. Even though we don't know how it works, most of the time it still gets the job done. That Alfa Romeo, by contrast, will definitely stop when its metal welds wear out or it smashes into a brick wall. Still, it made for a provocative slide that crystallized the difference in perspective between Héctor and the biologists.

THE SLIDE MAKES it seem like Héctor was hoping to simplify biology with physics, but in fact he wanted to do the opposite. While his fellow synthetic biologists wanted to make biology more LEGO-like, Héctor wanted to explore whether complexity and redundancy play a role we don't yet understand in biological systems. Why does the termite gut have so many genes that can do the same thing? In Phil's lab he proposed trying to build a bioreactor the size of a termite's gut before scaling that up. Then he wrote a paper with Phil and Katherine D. McMahon wondering whether synthetic biology should try to design its processes and output to take into account the diversity of natural systems.

Eventually, Héctor settled on measuring the flow of molecules among the tiny reactions within a cell. He spent a year and a half learning to do metabolic flux analysis, which involved growing microorganisms on food containing an isotope of carbon that is rarely found in nature, and then measuring and estimating how that carbon moved through the metabolic diagram.

In 2008 he joined JBEI to work applying his big theories to the relatively prosaic task of making some yeast or *E. coli* spit out some biofuel. He needed to track the flow of carbon atoms in these cellulose-breaking metabolic processes. By learning these fluxes and then making them intuitive with better metabolic diagrams, he hoped to eventually make them predictable.

Héctor's other self-imposed task was to design ways for the lab work to be more quantitative and systematic. For starters, he needed to get the scientists to upload their data to a centralized database instead of scribbling it in those black books. At an institutional level, combing through data from successful and failed experiments alike could give new insights. And even before he got them to upload their data, he needed to encourage the biologists to standardize their work at the bench. Doing molecular biology is like cooking: results seem to depend as much on the lab where the work was done as on the variables in the experiment. Venture capitalists estimate that only half of biology experiments can be reproduced. Héctor was dreading the task, because even at JBEI it was common in the lab to put mascots and good luck tokens on the machines. "Is this science or voodoo?" Héctor said. *That* word comes up a lot.

Héctor comically dramatized his frustrations with what he saw as the backward ideology of current biology, but he was sincere. These conflicts—which could seem abstract to a nonscientist—were deeply troubling and personal to him. He gave me a preprint of an essay that the physicists Carl Woese and Nigel Goldenfeld were about to publish in the *Annual Review of Condensed Matter Physics*, titled "Life Is Physics: Evolution as a Collective Phenomenon Far from Equilibrium." Héctor's biologist girlfriend became furious when she saw the title. "There's no discussing this rationally," Héctor said.

I read the article multiple times, though it was rough going. It called for a new understanding of evolution as a collective phenomenon involving time and ecology, rather than the individual stories of organisms and the survival of those fitting their niche. The authors said that evolution wasn't just about rewriting DNA codes, but a simultaneous writing and rewriting of the rules of evolution, like an Escher illustration of a hand drawing itself. They lamented the state of biology: "The lack of widespread appreciation for, and understanding of, the evolutionary process has arguably retarded the development of biology as a science, with disastrous consequences for its applications to medicine, ecology, and the global environment."

I was as interested in the ideas that made Héctor tick as I was in JBEI's quest to make biofuels. Over the next six years I visited him and his team many more times to see how they managed the process of working on small, slow, incremental research while their bigger, headier goals loomed over them.

AS THE GULF OIL SPILL dragged on through the summer of 2010, I wrote op-eds and traveled to Washington to talk with senators and congresspeople. I tried to explain that the problems of oil went beyond the failure of this individual well. Oil is more than a valuable fluid; it's a system that shifts costs, including pollution, risk, violence, and lost opportunities, off on others—including, for the moment, the human, avian, and aquatic residents of the Gulf of Mexico. Despite the opportunity provided by the disaster, lawmakers didn't enact any major policies to deal with oil consumption or climate change. Our oil and climate problems were deep and historic: we owned them. If this disaster couldn't shake our politics, perhaps synthetic biology, with its potential profits and lack of negative history, could change the discussion.

As I mulled this, I happened upon an article about the invention of nitrogen fertilizer that read like a historical case study of what happens when a technology solves a pressing problem and offers almost limitless growth—much as Keasling and Venter have described the promise of synthetic biology.

Until 1909, the planet's budget of bioavailable nitrogen was limited: we got by on the nitrogen compounds produced via lightning and bacteria. In Europe and North America, fields were spread with manure and crops were rotated to keep them fertile. In the 1800s, fertilizer arrived in the form of bird guano mined from islands off the coast of Peru, which contained plenty of nitrogen in the right form. In 1909, shortly after Einstein came up with his first theory of relativity, the chemist Fritz Haber figured out how to synthesize fertilizer from the nitrogen in air, giving us a supply of fertilizer for crops that was cheap and virtually unlimited. Because nitrogen compounds also happened to be essential ingredients for explosives, his process was quickly commercialized by the German military. At the end of World War I, Haber's process was such an obvious boon for society that the right to use it was mentioned in the Treaty of Versailles.

We often speak of the nuclear bomb as the defining invention of the twentieth century, but for ordinary people, the real change was nitrogen fertilizers. Especially after 1950, when fertilizer production ramped up to mass scales, our ability to grow food exploded, allowing the world's population to grow from 1.9 billion a century ago to 7.5 billion people today. On a molecular level, Haber's invention changed the planet, and it changed us, so that now about half the nitrogen in our bodies is derived from the Haber-Bosch process, which uses natural gas to produce nitrogen fertilizer.

Two things struck me about this story. First, though providing more food for more people was the ultimate in responsible innovation, it also had negative consequences. Nitrogen fertilizer arrived at just about the same time as gasoline-powered automobiles and airplanes. As the world's population began to grow, so did our use of fossil fuels, leading to our current issues with pollution and greenhouse gases.

There are some scholars who say that the invention of nitrogen fertilizer and the industrialization of agriculture have been a human-caused disaster for the planet. But that perspective is pure hindsight: I doubt anyone contemplating whether to make nitrogen fertilizer in 1910 could have seen any tension between the idea of saving more people and saving the world. And now, a hundred years later, we don't see a contradiction in solving our environmental problems with grassoline and unlimited hamburger. But our great-grandchildren may.

The second thing that struck me was something that seemed ironic at first: we once worked mightily to figure out how to use natural gas to make fertilizer to grow crops, and now we're laboring to do the opposite—turn plants into replacements for fossil fuels. The more I thought about this, the more I wondered if we focus too much on the inventions—the fertilizer, the biofuel, and even the wellhead itself—rather than the systems of power that grow around them.

THIRTEEN

RESTLESS STREAMS

THREE YEARS INTO MY TERMITE OBSESSION, they tunneled into my life. The front door of my apartment in Berkeley suddenly became hard to shut. At first I ignored the underlying problem by alternately yanking it open and kicking it closed, but finally I had to face the fact that something had happened to the shape of the doorframe.

My landlord called a carpenter, who entered my studio and pointed to the origin of the problem: the wall behind my bed. When he peeled the siding off, I could see that termites had chewed through the beams while I slept. What remained of the wood was fragile, lacy, and collapsing like old cardboard boxes left out in the rain. My bug roommates had left a set of shabby ruins for me, and here I was, living right on the edge of the Hayward Fault. A good shiver of an earthquake could have brought the house down on me while I slept. Termites had been chewing on my fate. All this time.

This revelation was eerie, and the fact that I saw the termites in my walls as interesting rather than horrible showed me how deep into the bugs' world I'd gone. What most people know about termites is that they eat houses, and yet I'd missed the obvious fact that they were eating mine. As the carpenters began their work, I had to admit that all of these insect "vacations" I'd been taking were more than just an escape from oil and work; they were giving me a purpose and some joy, too. It seemed logical to work out my obsession by writing a book about what I was seeing—of both termites and scientists—and it was a good cover story for this expensive hobby. But I didn't realize that termites were about to lead me somewhere else.

In late June 2010, a few days after I interviewed Héctor, my father called to say he had kidney cancer. Though we were close in constitution, we had not actually been close for years. When I was three, my dad trapped muskrats for extra money while working as a teacher. Every morning before dawn he'd put on his hip waders and walk his trapline, collecting muskrat corpses from swamps near the Connecticut River. I loved the way he'd return smelling of cold swamp, a musky kind of grease he used, and cigarettes. When I heard him come in the door, I'd get out of bed to talk to him.

In the evenings he'd sit at a high bench, skinning the muskrats and putting their pelts on stretchers, which hung in dozens from the ceiling of his workshop. I admired this so much that he built me a small bench so I could sit next to him while he worked. He'd cut off the muskrat feet and hand them to me, singing little nasal anxiety songs as he worked. The feet had horrible long yellow toenails still full of swamp muck. Sitting at the bench, I played with them as if they were dolls, making up conversations to amuse us both. Later he built very fine furniture, often with complex structures and dozens of joints that had to be assembled in order, rapidly and unerringly, before the glue dried. He had a terrible temper.

When I got back to Maine, he wasn't the person I had come to fear, but a frail man trying to navigate hospitals, Medicare, cancer, and mortality. I drove him to Boston so he could sign up for a vaccine study that would remove some of his immune cells, fuse them with cells from his tumor, and inject them back into him in an effort to make his immune system fight the tumors that had spread to his lungs.

My dad wasn't an obvious candidate for biological engineering. He'd spent a lot of his life rejecting the aspects of modernity that he didn't like. We lived as back-to-the-landers for more than a decade, using a horse to pull logs out of the woods to burn in our stove. And it was cold in central Maine. We raised and killed our own meat, baked our own bread, canned and froze our own food. Once, when my father learned that my mother had used store-bought canned pumpkin in a pie, he stopped eating and pushed himself away from the table. But now technology seemed like a fine gamble to take.

After each infusion he was supposed to inject his thigh in a pattern resembling a clock. Whether this was such a complex medicine that it required injection in a clock shape or whether we were involved in some sort of positive-thinking witchcraft I couldn't tell.

It didn't work. Sometimes when they cut out a kidney tumor all the satellite tumors go away, as though the original one had the powers of a termite queen. But not this time. Each tumor seemed to become an energized metabolism, uncontrolled by whatever normally controlled it. We began living in our own nonequilibrium state. His calcium levels started to rise, which indicated that osteoclasts were starting to unpack the calcium from his bones. Because Scott had compared the way termites build mounds to the way we build our bones, I imagined termites disassembling those white structures, leaving them fraught with holes.

There is a moment in *The Soul of the White Ant* where Eugène Marais describes a mound that panicked during a drought. "It took the form of a terrific onslaught, engaged in with such fury that the workers and soldiers could spare no moment for rest. It was a mighty struggle against death's stealthy approach; there was no respite for the defenders day or night." After dark the activity only got faster. "It was at night, during the hours when the rest of nature was quietest, that the fierceness of the fight gained most frenzy. I could hear distinctly the unceasing alarm calls of the soldiers, a sound which roused even in me a feeling of terrible anxiety. My electric searchlight revealed the restless stream constantly passing to and fro, as sure and indomitable as fate itself. . . . The death of a thousand individuals made not the least impression on that living stream. Vaguely and faintly, I began to realize, as I watched, what the struggle for existence really means in nature."

Of course, imagining termites was a story I told myself; clocks I was drawing on my own thigh. This was not science but the way that I was trying to process the information and feelings I had.

During our long drives to and from Boston my dad would sometimes enter a reverie where he talked about muskrats and their secret ways. Beavers and muskrats are great ecosystem engineers, and they've built much of the landscape of North America. Whenever we drove by a swamp, he'd talk about the muskrats it must contain, working to dam its watery whorls, capable of swimming under ice for a long time because of some special hemoglobin in their blood. When the hypercalcemia was particularly bad, making him feel ill and loopy, he told me stories about the muskrats, seeming to go off to their world. Sometimes he'd trace a spiral in the air with his finger as he talked about them. Now I wonder if he told those stories to amuse me and calm the frenzies I was feeling.

One day we walked slowly into a swamp he liked and sat on stumps to watch the muskrats we couldn't see. We sat there for a while, and finally it was time for me to take him home and drive to the airport to go back to California. But there in the swamp, all around us, was life: tearing up and tearing down.

That fall I could see the invisible architecture of the world within us and without us that is only exposed for brief moments in our lives. These are the moments when we can feel our own emergent selves, the gorgeous and improbable luck that all of my cells and all of my microbes happen to be heading in the same direction at the same time.

Marais was deferential toward dying mounds. One day he broke into the royal chamber and watched the soldiers shaking their heads and doing what he described as a dance around the queen. A piece of mud fell upon the queen, who began moving her head rhythmically. At first the workers wandered around aimlessly, and then they began to lick the queen, draining her of fluid. Work ceased all over. The workers collected in groups. Surprisingly, the queen seemed to recover. Marais returned the next day, "The queen was removed from her half-cell and taken away captive; and after that the activities and life of this nest ceased for good."

PART IV

FOURTEEN

CROSSING THE ABSTRACTION BARRIER

MASSACHUSETTS

I STAYED IN TOUCH with the roboticists Justin and Kirstin through email. In the fall of 2012, they mentioned that their TERMES robots were getting better at building walls, so while I was visiting Maine I drove down to their lab at Harvard's Wyss Institute. I wanted to see the robots, but I was really hoping to learn that building them had made the researchers see something new about termites.

The door to Radhika's office was draped in the kind of yellow caution tape you find at crash scenes, but when I got closer I could see it read: "Abstraction Barrier Do Not Cross." This is a computer programming joke. Programmers break down complex tasks into sets of abstractions: representing data, or manipulating it, for example. But these different sorts of abstractions must be kept separate—with barriers—or confusion will result and the programs won't work. I took it as a more metaphorical comment on the potential hazards of building complexity upon hierarchies of simplified abstractions.

I knocked on the door and Radhika answered. Her office was decorated invitingly with ant puzzles and colorful Escher-like paintings, but I didn't get to gawk because she quickly led me down the hall to the TERMES robots, which sat in the corner of an empty lab surrounded by stripes of black electrical tape. These robots were not lovely: three squat machines, each about the size of a Kleenex box, with pincers in front and beetle-like half-round humps on their backs. They didn't have legs, but things that looked like hooked ninja throwing stars. Near them stood piles of plastic blocks, each almost as big as the robots. Justin and Kirstin slouched against the wall like proud parents.

Kirstin clicked power buttons on the robots' underbellies and they began to waddle, an ungainly gait caused by their whegs—the star-shaped wheel-leg hybrids that allow them to climb steps. Their little motors whined with strain. Each robot began to follow a set of predetermined instructions: pick up a block, follow a path, climb onto the structure made of blocks, and drop the block. Their rules for placing the blocks are similar to the simple rules you follow when you're trying to avoid painting yourself into the corner of a room. In their case they work in a way that increases the height of the walls without trapping the robots on those walls.

People who care about self-organizing robotics were amazed by the TERMES because they were truly autonomous building robots: you could leave them alone all day and they'd build a four-sided castle. They were the first set of robots to answer a challenge posed in 1995 by Eric Bonabeau and Guy Théraulaz, two of the original swarming theorists, about building robots that could build structures—like wasps do—without supervision. But upon seeing them today Radhika seemed disappointed: "The other thing I'd love—in a year's time—is outdoor robots that pile up sandbags."

Kirstin knelt in the arena's grid. She was clearly happy and in her element. Officially, she was now a PhD candidate. She still spoke very quickly and under her breath. "They don't have a huge amount of memory," she said before gently placing a blue-whegged robot on the floor. Maybe she was soothing it.

"It doesn't take many rules," Justin interrupted. "You could write them all on a matchbook. And not even in very small letters."

TERMES has, to insult Descartes, a mind-body issue. Kirstin built the robots, embedding a kind of mechanical intelligence, a subrational reasoning in their motors and hinges. Justin built the algorithms, the robots' superegos, existential instructions that combine a flowchart and traffic rules. The blue robot began a slow limping trudge along a white piece of tape to find its way onto the blocks.

One of the robots got hung up placing a block behind us, spinning its whegs and whining. Kirstin glanced back to see if the robot was in trouble before explaining that the blocks contain magnets to help arrange their position, but the robot has to do a basic manipulation to get the block in the right place. On the third try the block clicked into position and the robot released it.

Maybe I looked a little bit surprised at its success, because Kirstin explained: "It does a sequence of actions that result in accuracy ninety percent of the time." I put a little star next to this in my notes because I wondered if the roboticists had invented the termites they had set out to find in nature.

Wherever possible Kirstin had replaced reasoning with reflexes, perfectness with probability. The prongs of the pincers automatically release when they're lowered, so the robot doesn't have to "remember" to release the block. The TERMES have ten sensors, including six infrared sensors that help them navigate and line up the stripes on the blocks, a tilt sensor, and push buttons that sense when the robot touches something. Additional ultrasound sensors help them identify walls and other robots. With this simple set of perceptions and three motors, they can navigate the world of the arena, performing their tasks. They even have a sort of electronic boredom: they'll repeat motions three times to try to get them right, but after that they give up.

Yet merely lifting and putting down blocks, no matter how accurately, wasn't enough. The TERMES team needed to prove that a swarm of independent robots could build a structure autonomously, without outside coordination.

So Justin wrote the algorithm, the set of universal sequential instructions to determine what actions the robots take, depending on their situation. Resembling a logic game like Towers of Hanoi, where you're only able to move one block at a time, the thinking behind such simple instructions is actually quite complex. The algorithm needs to explain everything—not just what the robot can and can't do, but what the traffic rules are. If a robot detects another robot ahead of it, it must determine whether the other robot is on the blocks. If so, then the first robot pauses until it stops sensing the other robot. If the other robot is not on the blocks, then the first robot goes to another program: "perform collision avoidance." A sequence of 106 steps over the course of twenty-four minutes led each robot to place blocks on walls. But no robot had a "plan," and the structure itself emerged from following these instructions.

By running the instruction set in a computer simulation, Justin determined how to tweak the instructions to get the robots to build a simple wall, an L-shaped wall, or a four-sided castle. In this way, he figured out how to reduce the "global" shape of the walls to a set of individual "local" instructions that were the same for each robot. Together the robots made a simple swarm from their sensors and actuators and algorithm.

And this is where the project got tricky. Building autonomously requires more than simply an algorithm that works on a computer and a functioning robot. Autonomous algorithms magnify slight unpredictability. So Kirstin created predictability by doggedly applying rationality: This was no place for intuition or educated guesses. To design the whegs, she 3-D printed and assembled fifteen different types of wheels, treads, and legs, evaluating each one for climbing ability, the angles they could sustain, and the smoothness of their gait. She did ten trials for each of the fifteen configurations. She refined the winners further through rigorous testing. She didn't believe anything but data, and to get better data she'd apply more abstraction. Reliability was the goal: everything else could be given up.

I have lazily thought of robots as "smart." But no one wants a robot that really thinks. We want reliable ones that can do complicated jobs thousands of times without any mistakes. For that robot you have to do a lot of unthinking.

One way to further abstract the robots' reasoning is to make their environment and even the blocks themselves smart. The white tape against the black floor gives them a path to "see" with their infrared sensors. The blocks, which can be 3-D printed or molded, are painted with stripes that help navigation, and they have little slots that help TERMES's whegs climb the blocks, and magnets to lock the blocks in place. On the top of each block is a sort of bowl that contains the robot when it's turning so its whegs don't fall off the block. In this way, the environment is also engineered and tuned to suit the robot's sensors, so that there is a sort of collective evolution between robot and niche.

This reliability carries with it a paradox: the more "reliable" the robots are in their arena, the less they can relate to the real world. A well-engineered robot will behave correctly in the arena it was designed for, but sometimes another room—where the room tone or lighting is different—will undo them. Take the stripes off the floor and they're lost.

Kirstin says that roboticists are constantly making excuses for their robots because they're part of what she calls the "toy world." "We're not even close to reality," she groused. "Our robots can only build on a black-and-white floor. They only work one of seven days. And a room can't have too many sounds, or the units won't be able to follow the walls."

Still, the TERMES gave a distinct and unnerving impression of being social as they whirred away behind us. The stereotypical robot follows the same sequence of motions over and over, something like a kid "playing robot." But the TERMES were playing the game as it lay—responding to the conditions in front of them. They really reminded me of toddlers doing parallel play, where they'll all appear to be ignoring one another while doing the same task, but are actually playing in relationship to one another. I mentioned that they seemed personable. Kirstin said that she sometimes left them toiling in the lab, only to run back down the hall, heart pounding, to turn them off—as though she had left children alone in the room.

Radhika looked at the TERMES, whose battery packs had run down, causing them to groan and limp back off their structure to the floor, and she seemed struck by both the limits of the robots and the limits of the problem she and her colleagues have set for themselves. "If it was a different world where they were really like termites, they could climb more and you would have to reason less. . . ." It made her want robots that could live outside the arena. "Seeing the termites climb up and down—we have to cross that, where the robots can fall off of structures. The world is just a whole lot less predictable. I don't really know how to do that."

It was around this point that I glanced up from the robots and noticed the poster hanging above the arena: it was a scanning electron micrograph of a termite's head that had been dipped in gold and then spattered with electrons. The magnified poster termite appeared to be made of metal, eyeless, with a head shaped like a dented medieval helmet and machined segments on its antennae. It didn't look natural at all. And the portrait loomed over us the way dictators' portraits hang over stadiums, spurring the roboticists toward the perfection of their impossible hybrids of bug and machine.

I wondered if the experience of failing to find the machine in the termites in Namibia had changed their minds about their work with the robots. Kirstin and Justin said no, emphatically. The TERMES, they said, were a short-term experiment, maybe another four months, and they had no further plans to explore termites. Really? I asked again, hoping that they'd show some regret about leaving behind the real termites. "Not even in all those videos you made?" They shook their heads. Kirstin put her hand fondly on one of the blocks and said, "We're probably going to abandon this project once they build a castle." On the other side of the arena the yellow-whegged robot stared at the block wall as though it had been given a time-out.

Seeing the roboticists' affection for their machines, I felt affection for them. I could almost understand why they were driven to treat the real termites as machines. After all, the roboticists seemed to interact with the world partly through their creations—both robots and programs. But I was selfishly upset that they didn't intend to do any more work with termites—either alive or as robots.

ON THE WAY BACK to Radhika's office we stopped at a glass case to admire an early version of the RoboBee, the Wyss Institute's famous creation. Though I had watched videos of it, I was surprised to see that it was only about the size of a quarter with its wings spread. It didn't look like a bee. It had a machine's aesthetic, with a rectangular frame for a body, and wings that reminded me of Leonardo da Vinci's sketches for flying machines. It lay immobilized in a caramel-size chunk of resin. This interment struck me as sad; it heightened the disparity between the ruined little machine and the great power of the idea of it.

Radhika seemed a little embarrassed by it, and I remembered something she'd said before: "Robots will only be part of our lives when we don't need PhDs to run them and they can make their own power."

Over lunch, Radhika mentioned, again, that she didn't want to save the world. She said this in response to a joke someone made, and it seemed as though she gave it no thought at all and was purely operating on a mechanical reflex: "No. I don't want to save the world." She shrugged—though whether she meant it's not possible to design an algorithm to save the world or that humans are beyond help, I didn't know. I liked her honesty. Reporters always want to hear about technology saving the world, and scientists too often indulge us.

But in the Wyss Institute there's no denying that technology does *change* the world. Kirstin pointed out three long glass panels displaying an old-style computer printout, with perforated sprocket holes on the sides and rows of dot-matrix printing. I assumed it was art, but she explained that it was a piece of the code written by Bill Gates, Paul Allen, and Monte Davidoff when Gates was a student at Harvard in 1975. The panels contained code for a computer called the Altair 8080—a funny metal box with some blinking lights that could only be programmed in machine language until Gates and Allen wrote code for an interpreter that allowed people to use the more accessible BASIC language. From this insight, Microsoft was born and the planet came to be salted with PCs and smartphones.

Now that we live comfortably settled upon this vast pyramid of abstractions and complexity, it's a shock to see the raw code underneath it all. The code that was hanging on the wall instructed the Altair 8080 to use a floating decimal to do math problems—a set of abstractions that allowed the machine to apportion its processing power, preventing jams. And because of all the abstractions that came after that, 1975 is a place—a way of understanding the world—that we'll never see again.

THE ROUTE TO a future of swarming robots will be built on greater and greater abstractions of both the equipment and the algorithms. Radhika mentioned that she had considered asking the team to build a robotic marching band, partly because they'd be forced to decide what to abstract and what to discard, but also because it would be fun. Now that I was at Wyss, I understood that the comic sensibility I'd seen in videos of their robots was a way to take the edge off the fear of failure.

TERMES gave a sense of how three fancy robots could work together, but the team wanted to show how a real swarm might act. A few years before, Radhika and the postdoc Mike Rubenstein had set out to build cheap robots for ten dollars. When they were done, they had the Kilobot, which cost fifteen dollars and took five minutes to assemble from electronic parts. These were two ends of the team's attempt to solve global-to-local: TERMES had a few robots doing complex tasks, while the Kilobots had hundreds of robots doing simple tasks.

We went to another lab, where six hundred Kilobots stood on a large table. Even less lovely than the TERMES, this robot is only a tiny puck of electronics perched on three stiff wire legs, with two variable-speed vibrating motors, an LED light, a battery, 32KB of memory, a little processor, and an antenna-like wire spring for charging. To go straight, a Kilobot switches on both vibrating motors and buzzes across the table. It turns by switching off one vibrating motor. The batteries can be charged en masse. Hundreds of Kilobots can be programmed at once in about thirty seconds via an infrared light over their table. They can "see" light and swivel toward it. They can communicate with the robots closest to them. When Mike turned them on, they started twirling and wandering around the table.

The Kilobots are goofy as individuals but sophisticated as a group, making something just short of a robotic superorganism. They can hunt by following a random search pattern; when they discover "electronic food," they communicate to one another until they form a bucket brigade to transfer the food to their nest. They can perform the mathematical averaging that allows them to synchronize their flashing LED lights in the same way that fireflies synchronize their lights.

As we watched, the Kilobots followed a lead robot around the table. A few fell behind—their sensors were off or one leg was slightly shorter than the others. I started to feel sorry for the struggling bots, feeling a kind of anthropomorphic pity for the 32KB machine. Hurry up, you're falling behind! I felt ridiculous for my inability to see them as mere motors. Justin nodded in sympathy, attributing it to the nature of the human brain. "Humans are wired to attribute emotional states to things. It's scared, excited, curious . . ."

With hundreds of identical robots there are problems—some break, or have faulty sensors—and the other robots must compensate. To show that the swarm could overcome these individual flaws,

Mike wanted to get them to build a self-assembling starfish. The starfish is an iconic figure in the self-assembly world because they are famously self-repairing: cut off an arm and they'll regrow it. In his dissertation, Mike had worked on self-repairing algorithms in computer models and now he hoped to translate these with a thousand Kilobots. Each of the little robots will need to know how to follow the edges of the robot swarm, understand their position relative to nearby bots, and be aware of a gradient, or distance, from the center. Following the edges of their group, the Kilobots begin assembling legs in which each robot has more connections toward the center of the starfish than away, eventually ending with a single robot on each point of the star.

Building a starfish would be nice—but Mike's ultimate dream was to build programmable materials. I asked what that meant. "Robots the size of sand. Let's say I'm camping and I need a number six Phillips-head screwdriver. I specify that, put my hand in the sand, and find it. It's a magic toolbox. You can make every tool you need."

IT WAS A DRIZZLY DAY in Cambridge, and things felt exceptionally leaden. But here we were, talking about self-assembling screwdrivers and magic toolboxes. And starfish for everyone!

The roboticists were dreaming of a great triumph of rationality: all those repetitive iterations and abstractions to find the perfect whegs and simulate the perfect interactions, all to create a collection of machines that together are emergent and fundamentally more than the sum of their parts, exceeding the rationality that built them. I understand why we want these machines—self-healing computer networks and the dream of applying swarm intelligence to the stock market—but I keep feeling like there's another paradox buried in the idea of gaining control by creating mechanical versions of systems that control themselves.

BY THE TIME I left the lab it was pouring and nearly rush hour. Boston's old streets and highways felt murky and unfamiliar, though I'd actually driven out of the city many times when my dad was getting cancer treatment there. Whether from rain or emotion I felt swaddled and blinded by the fog. I switched on my iPhone's navigation, hoping to get some clarity. Instead of steering me to my usual route, or even an alternate, it directed me to old Route 1, where I ended up in bumper-to-bumper traffic, the windows streaked with rain, listening to the early '90s hit by 4 Non Blondes: "What's Up"—with its terrible hiccupping chorus "Hey hey hey hey I said hey, what's going (hic) on?" Even though I changed stations, I heard that song at least three times before I escaped Boston.

Perhaps everyone around me had also turned on their phones, and we'd all been routed to the same place. We were the individual agents and we had built ourselves into a solid wall of cars while the classic rock algorithms of Boston had averaged to the ultimate anthem. I'd found the TERMES and she was me: sitting in traffic on Route 1, joining ten thousand other hapless termites. Our algorithm was optimized for each of us, and so it worked for none of us.

Maybe someday we'll run software that would optimize us as a group, correcting for errors and fuzziness, and we will go streaming along Route 1, or better yet, I-95, jamless. Further algorithms could banish any song that rhymes "institution" with "rev-oh-loooshun" from my life forever. Maybe that was the real benevolent promise of cracking the termite's code.

In the jam and through the rain and dark my thoughts kept returning to that screwdriver. I wondered if my discomfort with the idea was related to the unpleasant thought of plunging my hand into a bunch of semi-aware nano-chunks. I don't really trust other people to stay in their lane when I'm driving on the highway, even though I know their self-interest in avoiding a collision is at least equal to mine. But robotic chunks have no self-interest, and I would have a very hard time trusting a bunch of them not to take my hand off.

My thoughts drifted to a science fiction book from the late 1950s called *The Glass Bees*, written by Ernst Jünger. It's about a Steve Jobs–like inventor who has created a world of miniaturized robots—including bees—while running an entertainment empire. The narrator of the book is a man-out-of-time former cavalry officer, raised in the world of honor and bayonets before World War I, who visits the magnate's garden in search of a job. In the garden, the cavalry officer sees a swarm of glass robotic bees devouring what appear to be severed human ears. He realizes that with the appearance of the tiny mechanical bees, all the morals and rules he grew up with have changed. The story stayed with me because the experience of the cavalry officer was precisely what happened in real life as Europe's military elite experienced machine guns, bombs, and poison gas turned on them in World War I. And yet, the book also saw a similar, or bigger, transformation in power and morals occurring with the arrival of its robot bees. One sentence stuck with me: "In the automaton, abstract power becomes concrete . . ."

I don't think there's anything inherently wrong with mechanical bees or self-assembling screwdrivers, just as there's nothing wrong with computer operating systems and cell-phone navigation. But, like the machine gun, automata allow us to abstract power in new and convenient ways that sidestep our existing systems of morality.

Already drones are permitting these abstractions. Just before dawn on October 10, 2012—the same day I sat in traffic—a U.S. drone strike killed five people in Pakistan. I had no idea then: we were not at war with Pakistan, and the strikes were not covered in the news here. But these strikes were very real in Pakistan, where thousands of people had driven their cars in a rally against them, led by the cricket superstar Imran Khan, on the previous weekend. Secrecy—or lack of awareness—was just one aspect of the abstraction of power the drones offered to Americans. Another was the ability to kill individuals without risking American lives—the operators of the drones were safely in trailers outside the action. As with the early days of the machine gun, drones were only being used in special zones that were not in Europe and the Americas—leaving open the question of whether these weapons would ever be used against us, and whether that should play into our thoughts about the morality of using them. And the unmanned drones were increasingly blurring the line between humans and machines, clearing space for autonomous drones to come into the mix, and making abstractions of insects, machines, and power all the more concrete.

FIFTEEN

INFLUENTIAL INDIVIDUALS

DESPITE HER INTENTIONS, Kirstin did not give up on real termites. For two years she kept returning to those unimpressive videos she'd made in the final days in Namibia in 2012. She looked at the petri dishes, half filled with mound soil and half fresh, and the termites wandering around and digging in them. These were their least impressive experiments, so low-tech they were hardly even experiments at all, but they bothered her, and so she decided to figure out how to track the individual termites. What she discovered was that the bugs did not act like robots at all!

In early 2014, she told me that her work had revealed that termites were individuals, with personalities. Or as she put it: "I'm sure there are key individuals who trigger everything." This was no hunch; only a roboticist would use the word "sure" in that way.

The very idea of termites with the kind of charisma that could "trigger everything" surprised and thrilled me. I had given up hoping that the roboticists would help me get to know the bugs better. Kirstin's new insights reminded me of the sensation I'd had watching Scott's videos of the green glowing termites exchanging fluids: Were they moving water, or was moving water a by-product of some more pleasurable interaction? Did it matter that we couldn't tell the difference? In Kirstin's ugly petri dishes, a paradigm had shifted and it wasn't clear where it would lead—for either termites or robots.

I took the train south to Boston to see Kirstin. By this time I had moved from California back to Maine, drawn there by obsessions (love, a river) not that different from those that drew me to termites. In the meantime, I'd tried forgetting about termites, but it hadn't worked. I thought about them all the time. So in 2013 I decided to make it official and write a book.

When I found her in Cambridge, Kirstin was wearing an enormous rigid wool sweater with a neck that covered her chin—part armor against the bitter cold, part exoskeleton. She was less reserved than before, but matter-of-fact about her success—or the previous failure—depending on how you looked at it. "You never get what you're looking for in biology. We thought of every termite as the same termite. We thought every colony was the same colony. We spent a year doing nothing. We didn't find what we were looking for."

She had a little plastic box of live American termites on her desk. As she spoke with me she sometimes opened the lid and peered inside, smiling, as though she was discovering a pleasing miniature world. Now that she'd invented a way to track the termites, she couldn't believe the team had considered—for even a minute—that they were identical. Even robots aren't identical, she said. "We were idiots. When I build robots, I know they are not the same. It's laughable. Every robot has a different personality."

Kirstin asked two undergraduates to work with her on trying to track the termites in the dishes. They had to do this by hand, at first, because off-the-shelf bug-tracking software didn't work for termites. (Ants are often painted for tracking, but termites will groom the paint off one another. And another program for fruit flies tended to lose track of individual termites when they crossed paths.) Every day, Kirstin and an undergrad would track one of the twenty-five termites in the dish through all 999 frames of video. The whole process took more than two hundred hours. The human workers didn't just identify where the termites went in the dishes; they also identified when the termites were digging, when they were carrying a mud ball, when they dropped it, and when they returned to a place they'd been in previously. This was the first time such a big set of high-quality data had been gathered for a dish of termites, anywhere.

For serious termite tracking, this was ridiculous, so Kirstin designed some new software using filters and neural networks, but it was not as accurate as human trackers. Things improved when a grad student figured out a simpler way to do the tracking that used Bayesian statistics. The program drew an imaginary circle around each termite's head and used a statistical program to predict the probability of where the termite would go with every step. It averaged the data and then applied a gradient and compared the likely outcome with what appeared in the next frame of the video. Because she had the hand-coded data from the same dishes, Kirstin could rigorously check the new software until it was more accurate than a human. Then the team made another program that could semiautomatically determine what the termite was doing at every moment: wandering, digging, carrying soil, putting it down, interacting with other termites, or resting. A researcher would still need to check the data, but after 150 million years of working as an obscure mass in the dark, termites could now be seen and tracked as individuals.

Kirstin analyzed the data in hundreds of maps, bar graphs, scatter plots, histograms, and frequency charts. The graphs were colorful and obscure. I stared at them for a long time before I understood what they meant: If termites were actually factory workers, most of them would be fired. Commentary below sequence 130716 noted that only five out of twenty-five termites were building. In another dish two termites did the building while four helped a little and the remaining nineteen just ran around. Kirstin said that when she started tracking what each termite was doing—not just where it was going—she discovered that even though some ran around a lot, only a few made progress on the actual building. Termites seemed to do whatever they felt like: dig, take up soil and clean the dish, sit around.

Kirstin's termites were not the downtrodden drones of a totalitarian assembly line, but something more like the well-kept residents of a Danish socialist village—each contributing in its own way. This was really something new in termite land, but it has been observed in ants. When researchers in Anna Dornhaus's lab at the University of Arizona tracked 250 ants from five different colonies, they found that about half of the ants do nothing; they do "nothing" so thoroughly that it could be a job description. The researchers speculated that they may be a reserve fighting force, snack bearers offering some trophallactic refreshment to their colleagues, or they could be guards, "data hubs," old, or young . . . No one proposed that they were philosophers or Bill Murray characters, but it strikes me as funny how hard it is to suggest that insects are slackers when both evolutionary lore and Aesop's Fables agree that social insects must be industrious.

When Kirstin checked the videos, she'd do ten thousand frames an hour. I watched the video as twenty-five termites wandered around the dish, their bodies blurry, segmented, more like collections of gray bubbles on a gray background than individuals. Termite One hardly did anything. He or she stood in the middle of the petri dish on the boundary between the clean soil and the mound soil. "This guy's easy. He's not moving, but he's clearly not catatonic." She advanced the video frames quickly. The termite's head bubble bobbed. "He's clearly smelling the dirt."

Termite Two ran around madly from one clump of termites to another, picking up nothing. Termite Three had something in its mouth. It put it down, wiggling its head. It ran to the side and then picked up another dirt ball and staggered irregularly back to the center line, where it found a spot and put the ball very close to the boundary line between the two kinds of dirt. It then ran back to the edge, nosing in with a group of termites who'd been digging furiously at the very edge of the dish, their butts wiggling so they looked like little piggies at a trough. Termite Three emerged with a ball and ran back to the border, where it dropped it. It backed away from the ball it placed, probably to avoid disturbing it. Kirstin said, "This supports my theory. I try not to be biased, but you can't help it. They're more careful about going backward when they have something in their mouths. See the difference?" She played the few seconds of video again, showing the termite backing up. "This is a great termite. It's actually supporting my theory." Termite Three was up to something.

Termite Four did absolutely nothing; it didn't even move. Termite Five played in the dirt. Termite Six ran with the other termites but didn't carry a single dirt ball. Termite Seven ran around and around. Termite Eight picked up other termites' balls and moved them a few steps. Termite Nine ran in and out of the crowd but accomplished nothing.

There were sixteen other termites in the dish, and we have only observed three minutes of their lives. But even in this short period of time, I can see that some termites have a different sense of their world and of themselves. There is an existential element to staring at these fuzzy videos; I'm forced to reappraise who the termites might be, and what that might mean for them, for nature, and for robots. Kirstin, having run her numbers backward and forward, has accepted it: "The informed individuals have a purpose. They have an opinion."

This kind of leadership has been seen in other animal collectives. Iain Couzin has done extensive work on how fish, birds, and ants self-organize, finding that "informed individuals" influence the way schools of fish swim, effectively leading them. When more than one informed fish begins to lead, the school will often average between them. Interestingly, the more fish there are in a school, the fewer leaders they need. The fish behind the leader have behaviors that allow them to concentrate their sensory awareness while damping down unnecessary hair-trigger reactions. This mixture of sensing and damping makes the system sturdy and does not depend on individuals or single leaders, yet it takes advantage of their enhanced abilities. Fish, in other words, have evolved collective strategies that reduce the likelihood of following a really eccentric fish with a bad idea, which is something humans might want to look into.

I watched Kirstin watch termites for four hours, just a tiny fraction of the time she'd spent on the project. Watching people watch termites—whether it was Scott sorting them, Anna smashing them with her pipette, or Phil playing jazz—gave me insight into the daily routine of doing science. It was boring, risky, lonely, and cerebral. And, where termites were concerned, undeniably trippy. As I'd noticed when I was at the bench gutting termites with Shaomei, to do repetitive tasks you really have to mind your head and stick to the details. For example, when working on the video, it was all too easy to make up stories about why the termites were doing what they did. As she was clicking, Kirstin muttered, half in warning to herself: "There's no such thing as intuition."

But Kirstin's data revealed a world that was more intuitive—more gooey, more individual, and less robotic—than the more mechanistic views of termites that humans had been able to imagine. It was as if scientists had forced themselves to obey a set of rules about how to think about what termites do—their own internal algorithm of possibility—and that led them astray.

ONCE KIRSTIN STOPPED assuming that the termites were identical to one another, all other assumptions came under strain. She had started to wonder about colonies: perhaps each mound has a different personality, or different genetic mix of personalities. A year earlier she'd been ready to give up; now she was having much bigger thoughts. She even wondered if she could influence termites to become leaders, perhaps wave a stick containing a pheromone in front of a few termites so they'd assume the role of directors. "I'm a firm believer in social influence over genetic," she said. And then she started wondering about designing small robots to interact with the termites to influence them to build different mounds.

In the lab as a whole, the idea that termites were stochastic beings had melted away, replaced by an appreciation of their individuality. Justin and Radhika applied for a prestigious grant from the National Institutes of Health to return to Namibia. They described studying how termites organize themselves as "useful not only as a matter of theoretical interest but . . . [applicable] to the functioning of human institutions."

Justin's office was a small room with a monitor surrounded by a collection of Muppets, African art, puzzles, juggling balls, and three sheep brains in jars. News about the termites had excited him, and caused a bit of reflection about where the group had gone wrong. Recently he'd been recorded telling an audience about his life studying termites. When he first started, he said in the story, he expected to watch termites for a week and be done. "Physicists believe they can walk into any situation and solve it from first principles. Physics is the Incredible Hulk. Physics Smash!"

Now he had new ideas about how to use the termites to further explore global-to-local. He explained that the question wasn't which theories were governing everything in a literal sense, but what the connections were between the little systems and the big ones. Some theoretical physics relies on postulates that are crazy, he explained, but it succeeds in describing actual observations. "Real string theorists don't care about observation, they're saying 'Here's the language of god that underlies what we experience,'" he said, echoing Héctor García Martín's observations on applying ideas from condensed matter physics to biology. If the team can connect the little systems among the termites and the big effects they create, tracking the termites might be like eavesdropping on god.

When I asked Scott what he thought, he nearly crowed, "Idiosyncratic individuals are driving the whole system!" It validated some of his big questions about how to approach biology's complexity: "The mechanistic view of life doesn't have room for minded things." And yet, it was machines that had revealed this "minded" world.

Kirstin's tracker's biggest fan was Scott's postdoc, the entomologist Paul Bardunias. A giant man in his forties, Paul has a wide-open face, like a Campbell's Soup kid. After taking time off from a science career to take care of his family, he got his doctorate at the University of Florida, studying how termites dig tunnels. When he's not with termites, he's writing about ancient Greek warfare—another subject on which he is an expert. This didn't surprise me at all: if you can get hooked on termites, there's always room for another monkey on your back.

Paul asked Kirstin to use her tracker data to test other things he'd wondered about. For example, do termites turn at a certain angle after they pick up a ball of mud? Kirstin spent five minutes writing a program and showed him that, yes, the termites did tend to head off at an angle of 30 degrees after picking up a mud ball.

While Kirstin pulled insights out of the dataset, Paul saw his job as finding the story in there. "It was like having a third lobe to my brain," he said. Simply having the data enabled a different kind of science, one where hypotheses could be tried in the data rather than in the experiments. This was similar to Phil's jazz, gathering data first and asking questions later, but it felt uncomfortable to Paul. He told me the tracker was "Arthur C. Clarke." When I asked him what he meant, he supplied a quote from the British science fiction writer: "Any sufficiently advanced technology is indistinguishable from magic." "It's beyond me. I'm a caveman. I can use it. But I can't reproduce it," he said.

Paul saw another important idea in the data: while he expected termites to drop their dirt balls on old mound soil, they also seemed to pick up balls from that soil. For Paul, this was a eureka moment. If the old mound soil contained a cement pheromone, then it should work like a key fitting into a lock, releasing exactly one behavior. But once you could see individual termites in the video, you could see that they did all sorts of things when they encountered the mound soil containing its possible pheromones. In fact, whatever they were doing, they changed it. If they were carrying, they dropped. If they were empty, they picked up. "It causes everything!" Paul explained. Technically, it appeared that the mound soil contained an arrestant that signaled the termites to finish up whatever they were doing. Paul called it a "Shalom" chemical, appropriate for any and all occasions, its meaning dependent on the context.

In fact, this was not the first time the researchers had seen this. When Justin put the termites in his Plexiglas bridge in Namibia and they formed a whirling 3-D scrimmage right on the boundary between the two soils, he was seeing the "Shalom" effect in action. The termites didn't want to leave the smell of the mound soil. But because the research team members were looking for other behaviors, they thought it was a mistake. Now that they understood what was going on, however, they'd have to reconsider other theories that were based on the concept of the cement pheromone—particularly stigmergy. Derived from Grassé's work in 1959, stigmergy had spawned so many thousands of hours of computer simulations and so much productive thought. It was the very thing that had drawn Justin into the study of termites in the first place. But now it would need to be reexamined. He was philosophical, "You don't get to choose what you don't understand."

Over the next three years, Justin (who had gotten that grant from the NIH) would lead the team into a new understanding of how termites build. The tracker was refined and rebuilt. In Namibia, Paul and Ben Green, a grad student in applied math, carefully gathered the two species of *Macrotermes*—the ones Scott had speculated had longer attention spans to build tall mounds, and those who built shorter mounds—and began doing experiments with them. The results replicated the broad outlines of Kirstin's: the preponderance of termite slackers, the few who influence the others, and the mound dirt that smells like home.

But they also came to understand that the cue for building—like the sound of running water for beavers—was digging itself. The concept of stigmergy, in other words, might be upside down: instead of being driven by dirt balls that inspired further dirt balls, it was driven by digging. When a few termite individuals started digging, others would join them, shoving in—as we'd seen—like pigs at a trough. Every additional termite jammed into the digging pit increased the odds of the next termite joining in, until there were just a few pits filled with lots of termites. This frenzy of digging resulted in the production of dirt balls, and those got stacked in little spires, walls, and arches that eventually became solid. Somehow, digging created a template for building—and the researchers were able to confirm this with computer models. What's more, termites had some sort of memory, because their past actions influenced their behavior.

The new ideas about termites led to new robots. Paul started wondering whether the termites were digging at places with high humidity, and whether part of the attraction to other digging termites was the humidity around their bodies. Since the humidity in the nest was over 92 percent, driven partly by the fungus's evaporation and partly by water in the environment, water vapor was a defining feature of the mound, dividing the inside from the outside while uniting the termites and the fungus. With Paul's work as an inspiration, Justin and the team of roboticists began to experiment with dribbling robots that can trace one another's paths and build along a moisture gradient. Muddy, dribbling robots seemed like something Radhika would like.

ALL OF THIS was far in the future when I visited the lab in winter 2014. Before I caught the train back to Maine, Kirstin offered to show me the TERMES robots again. In the lab they still sat under the poster of the big, dented termite's head. The little TERMES were outwardly unchanged, but they had recently crossed a threshold of reliability. One morning Kirstin had turned them on at 8:00 a.m. and watched them do four thousand alignments correctly, building their little walls. By the end of the day they had the dependability and low error rate they needed to be taken seriously.

Kirstin's competence of two summers ago had turned into amused ferocity, a productive engine. She was set to complete her PhD in under two years. She had made Radhika's edict of fun and difficulty her own. She was preparing to leave Cambridge to return to Europe, where she'd work at a new lab.

And during her time watching videos of termites she had gotten new ideas about the kinds of robots she wanted to build. The termites had changed the way she thought about machines. "I've reached the mindset that's much closer to the idea of constructing a society than three robots who can make stuff together." She was interested in building groups of robots who depend upon one another and have overlapping skills. For example, she was thinking of using very cheap sensors to create some robots who perceive more than others. "I'd like to make a lot of really dumb robots. Whoever gets the sensory information becomes the leader. Over time they will forget that they're dumb."

Current robots rely on perfection, with cameras and trackers, but what she wants is the ability to be sloppy, to be good enough, to be a system of individuals and reflexes that work right "at the edge of reasoning." She doesn't want robots that think, only robots that react, with a high tolerance for failure. "You have to design for crummy. Then it's a robust system. It would be great to move to a world that works ninety out of a hundred times."

Standing beside Kirstin, looking down at the little TERMES not much bigger than our shoes, I thought about the day that the robots would leave her toy world and join our real one. I asked her if she thought we'd ever see robots building on Mars, and she said she hoped to see them building on Earth in our lifetime. It's not inevitable, but the relentless application of abstraction eventually does get results.

Kirstin's view of the process of progress was termite-like: as more people got interested in building autonomous robots, they'd bring their own personalities and methods to the hard problems, and then the field would start to move. Carnot might not even be necessary. As I thought about her description, I realized that stigmergy might not be an accurate way to describe termites' building, but it seemed like a decent metaphor for the progress of scientific ideas: ideas that smell nice draw more scientists with different personalities, and eventually—the way mud balls adhere to become a mound—they build something real.

And then I thought about these new robots themselves: someday there will be billions of them, as different as snowflakes, cooperating and goofing off as mobs of individuals. A million mechanical Three Stooges will create real buildings as slapstick comedy. And when this social machine, this deconstructed brain on whegs, is finally among us, most of us will be cavemen, understanding it mainly as magic: "Arthur C. Clarke."

Kirstin smiled down at the ugly little TERMES and whispered, "These beautiful flocks of droids."

SIXTEEN

THE ROBOT APOCALYPSE

KIRSTIN, JUSTIN, AND RADHIKA submitted a paper about their TERMES robots to *Science*, which decided to put the TERMES robots on the cover of the February 14, 2014, issue. At the same time, the American Association for the Advancement of Science asked them to make their robots "perform" all day at a conference in Chicago. TERMES were going to be famous, but taking the robots out of their arena—with its particular lighting, sound, and electrical tape—was tempting fate. So Kirstin went along, and when the TERMES's batteries froze in the Chicago cold, she was there to fix them. When someone dropped one, she fixed that too—even after she accidentally ripped the skin off her palm with superglue. Radhika's observation that robots won't be real until they don't need PhDs for repairs seemed about right.

But frailty and unpredictability were not what the world saw when they looked at TERMES. One news report talked about "colonizing Mars," while another deemed robots the future of construction, adding, "Let's just hope those termite-style robots don't become self-aware and cause carnage." These imaginary robots were all-powerful. "Has the robot apocalypse arrived?" asked a video on *Bloomberg News*. Mixing footage of TERMES with old horror films, the video attributed near-magical powers to their algorithm, ending with the promise of building on Mars and rescuing humans before showing black-and-white film of a lightning strike with the reassuring words, "Not yet."

I don't think the roboticists were surprised by the tone of the reception. While synthetic biologists are presumed to be saving the world, roboticists nearly always have to endure some kind of Frankenstory. Science redefines itself almost daily, but public discussion of the morality and possible outcome of technology remains stubbornly anchored to Mary Shelley's story from 1818—six years before Carnot wrote his theory.

THE ROBOTICISTS didn't think much of the robot apocalypse as a story. One day over lunch, I'd listened as Radhika's team discussed Michael Crichton's 2002 bestseller *Prey*, about a swarming nanotech project that escapes from a government lab to terrorize the world. Radhika said she liked the parts of the book that dealt with self-assembly. For example, at one point the swarm of nanobots makes itself into a pinhole camera so that it can see. "I thought of proposing that we build that," she said. She thought the second half of the book was terrible because it dropped the technology when the nanobots invaded humans and made them into horrible, almost supernatural beings.

Justin agreed. He had read the book hoping for a thoughtful discussion about swarm intelligence and self-assembly, but he thought the story was an outrage because it shifted the whole narrative away from the very real potential of the technology to the more religious concept of human evil.

"Oh, so it turned from a swarm story into a zombie one?" said Nils.

"Yeah, and the world ends," Radhika said. Uninteresting, unreplicable, and empty: the battle between good and evil was the living dead of discussion topics for them. They would have enjoyed a conversation about complex systems, a weighing of the trade-offs between reason embedded in hardware and superego in the software, or a philosophical take on building emergent swarms with flawed machines. But they didn't get it.

"Bummer," someone said.

IN MAY 2014, a few months after the successful launch of the TERMES, I returned to Cambridge to interview Radhika. Before we started I had time to examine the paintings hanging in her office, which Radhika had painted. In one, a large red-and-white tree dissolved upward into birds, bees, termites, and flowers, which then condensed to fleurs-de-lis and abstract shapes, maybe even some RoboBees. The painting seemed to be a commentary about her work in biology and robotics, a visual depiction of emergence and self-organization. But after a minute or so I saw a woman's back and hair entangled in the tree. It was not revealing, as a self-portrait might be, and so I did not immediately assume it was Radhika herself, but more of a generic Earth Mother creature, or maybe a pink tree lady. I was surprised when I saw that the painting's name was "Losing Myself." To my eyes the person in the painting was anything but lost: she was positively embedded.

There are very few people who've gone as deeply as Radhika has into the crack between biology and engineering. She had both the closeness and the skepticism to understand how swarming robots might change us and the world. I started with a question that seemed obvious to me: Why had she said, twice, that she didn't want to save the world?

She seemed surprised by the question but agreed that she said that often. Her first objection to the idea of saving the world was the mixture of pressure and hubris involved. "Really, I'd have to have an arrogant view of myself." Secondly, most of the things that will save the world have nothing to do with her research. "None of the robots I'm doing will deal with climate change. And if climate change goes to the worst, our lives will be over and there will be no need for these robots. So clearly, if I wanted to save the world, I'd stop doing what I'm doing and work on climate change. Energy. Or poverty."

I asked about the painting of the red tree, hoping to move into the emergent part of it. Radhika said she'd painted it while she was on an interdisciplinary fellowship that allowed her to hang out with historians, biographers, literature professors, and fiction writers. "I'm now used to understanding biologists, but I don't understand people who study literature at all," she said. "They all have a different value system that drives what they do, and they don't understand mine and I don't understand theirs."

Because the other fellows wanted to know how social insects related to utopias, Radhika read up on utopias for the first time. I thought this was funny because when I was twelve or so—the age when she was living in Amritsar—I was bookish and got some of my clothes from an old trunk that an uncle had left behind. I escaped from the disaster of junior high by imagining little self-serving utopias where everyone wore blue uniforms—a sort of termite colony where I finally fit in.

Radhika's interest in utopias was of course less conventional. The usual debate about utopias is the flip side of the apocalypse: If human evil doesn't bring us down, can human good bring us up? In a more academic way, utopias mirror the discussion about the role of altruism in evolution; the consensus is that doing things for the general good could never work. Radhika didn't think of utopias that way at all: she thought they were an algorithm, a set of rules—human good and evil had little to do with it. What she wanted to know was what you'd get if everyone *did* follow the rules. Would *that* utopia actually be any good?

Her question was far more interesting, and what it revealed was how the discipline of science, practiced at a high level, is about open questions, multiple variables, and ways of addressing uncertainty.

Popular culture rushes science toward a happy ending (like saving the world) or a sad one (like an apocalypse), but these inherently conservative narratives, always calling up old stories and fears like there's nothing new under the sun, are diametrically opposed to the world the scientists live in. If we could move into that gulf between us and the scientists, and be content with uncertainty, I think we could start to think more deeply about what we're doing with the technology scientists are building.

Radhika's most lively painting was *Lost in the Dollhouse, Self-Portrait of a Modern Woman*. Resembling a cross-section of a termite mound, it's a portrait of a dollhouse that she painted in her third and fourth years of assistant professorship, when her kids were still little. In one room in the lower right corner her kids sat in their playroom, and in the uppermost compartment Radhika stared out the window at some clouds. When her kids were young, it was a dream of hers just to have a few minutes to look out the window. Despite its rich colors, I feel sad looking at the painting. "It was a hard year. That painting is about being overwhelmed and trying to compartmentalize. I felt like one compartment was leaking into another."

The compartment for research has her sitting by a stream flowing with molecules and creatures representing cellular automata and bio-inspired robots. It's the most introspective and absorbing of the compartments, and it reminded me of the garden in *The Glass Bees*. In her art Radhika crossed back over the abstraction barriers between this intensely intellectual work and the demands of teaching, motherhood, and being a person. In the dollhouse's kitchen, Radhika and her husband, Quentin, stood back to back, juggling vegetables and receipts between them. She pointed out that they were holding hands and juggling with their free hands. "It was us against the world. Our kids want ever more and our work wants ever more. And somehow we had to preserve sanity. At the end of the day, we didn't think they deserved that much of us."

In the previous year, Radhika had written an essay in *Scientific American* that went viral. Typically Nagpalian, it was an unconventional reinterpretation of academic requirements. She had taken a hard look at the "rules" for success and decided that most of them were irrelevant, and then she had tried to figure out what was fun and important enough to stick with it over the long haul. She ended up treating her career as she would a mechanical system—abstracting to the most important functions and then cutting features until she had a decently robust system. And, like her robots, work had to be fun.

I had started asking her questions about the paintings because I wanted to get a sense of how she saw robotic swarms entering our world. I hadn't intended to get into her family life. It's unfair to ask women scientists about their families and their work when reporters mostly ask men about their work. But Radhika had recently decided that the system needed to be changed to remove the unspoken threat against having children. "If you don't want me to have children, you need to put it in the contract that I need to be celibate, otherwise you don't get to treat me like I'm abnormal. You can't have it both ways where you say one thing and do another," she said.

At that moment I heard a sound outside the room that was simultaneously strange and very familiar. There were male voices, female voices, and a sensation of swaying. People were singing and seemed to be heading toward our door! Even without hearing the words, I knew that whatever was going on in the hallway *was not serious*. And so far, everything I'd experienced at the Wyss Institute was serious—even when Radhika was being silly, she was advancing big ideas.

The sound stopped just past the door. A man began exhorting someone—how many were there?—in a churchy cadence. I caught a few words: "Your soul is coming for us." "Misguided science." "We know we need the honeybee to live." A chorus sang: "Can't pollinate me. My bee is alive. Woowoo who woo whoa. Who wo wo wo bee."

Radhika said quietly, "I'm actually deathly scared of them. I'm afraid I'll get lynched." She walked to a cabinet, found a piece of copy paper, and handed it to me. I saw the words "Monsanto" and "Reverend Billy and the Church of Stop Shopping." I recognized Reverend Billy: he is a well-known performance artist who has crusaded against consumerism. I'd always felt warmly toward him. Last I remembered he was handcuffed to Mickey Mouse in front of the Disney Store in Times Square. What was he doing here?

"They're out there. I'm just afraid to go out. I have no idea what they'll say to me." Radhika switched off the lights in the office. She did not look scared to me.

"They're all dressed up," she said. "If you want to see them, go look." I did want to see them. I'm always up for a show, the weirder the better. As a reporter, I may be constitutionally closer to performance artists than to scientists. But I chose to stay in the room with Radhika because I wanted to know what it felt like to her. What is it like to be in *this* room with *these* lights off and the "Abstraction Barrier: Do Not Cross" in place? Who was the abstraction barrier for? I moved a little closer to the door so I could hear better. Radhika thought I was going outside and she said, "They'll think you're me." I turned around and sat down.

They had come for the RoboBee, that paper-clip-size bit of machinery lying in state in the glass case in the hallway. I looked at the paper Radhika had handed me. "They say you're funded by Monsanto," I said. The absurdist fog started to clear. Monsanto is a former chemical company that now produces genetically modified seeds and herbicides. It's a huge player in industrial agriculture and has many critics. I remembered the video of the young researcher talking about using the bee to pollinate crops. But that was just silly, because the RoboBee's most likely eventual suitor would be the military. While it was true that honeybee numbers had fallen dramatically, no one was crazy enough to pay for robotic bees to pollinate flowers.

"We don't get funding from Monsanto," said Radhika. "The fact is, the bees can't even carry a battery!" She emphasized the *b*'s in that sentence. "There's zero chance in anything like the near future. It's like the termite robots taking over the construction industry!"

So a message about the RoboBee had gotten out—but it was the Robot Bee Apocalypse. It crossed my mind that the roboticists, with their love of precision and abstractions, were the people least likely to enjoy performance art. Especially factually sloppy performance art. I felt embarrassed, as though I had something to do with the people performing in the hall. They were making us non-roboticists seem stupid.

I watched a video of the performance later. A tall woman they called the Queen Bee wore her hair in a beehive smothered in black netting decorated with large fuzzy bee effigies, and a man with a beard dyed caution-tape yellow wore a bee-encrusted top hat. They placed exotic fruits at the foot of the glass case where the ruined RoboBee sat in its square of resin. Reverend Billy wore a pompadour, a white suit, and a black collar. He preached with comic fervor into a megaphone: "You don't have to make a RoboBee and replace this bee that is dying. We ask you to place your genius, your artwork, your research, your scientific know-how in saving the honeybee because we cannot live on this Earth without other life. . . . We expect you now to start SCALING DOWN the RoboBee."

This was more than Radhika could stand—the RoboBee project had been going on for six years already. She tuned out the racket and returned to serious topics. Any suggestion that the RoboBee was doing agriculture was a cover story. Early on there was criticism that the government shouldn't fund trivial ideas like robotic bees. Then there was criticism of the potential military purposes of the bee. So the researchers came up with other possible uses for the bee, other stories really—like search and rescue and agriculture—for a technology that was twenty years away.

And, she added, watching the RoboBee on video didn't give an accurate sense of what it could do. It was confined to its arena and it wasn't so much flying as it was moving its wing flaps fast enough to keep it from falling. "And that's a big deal, because nobody knew how to build something with wings that could lift its own weight." The most significant outcome of the RoboBee, she said, was that it demonstrated a way to build very small devices using 3-D printing and origami-like folding to assemble them.

The role of biology in her field, Radhika said, was not really about imitating nature; they used nature as evidence that a hard problem could be solved, much the way Phil's team exalted the termite because its gut was capable of breaking down cellulose. "Here's proof. It can be done. If I haven't done it yet, it's because I'm stupid, it's not because it cannot be done. If biology with ten neurons can do this and we can't do it with a computer, then it has to be our mistake. We don't know how to think."

I asked her how long it would take for us to learn to think. When will we have swarming robots among us? Radhika compared it to the first airplane at Kitty Hawk—a proof of concept—that took decades to become a jumbo jet. It took twenty years to get from Kitty Hawk in 1903 to nonstop flights across the United States, aircraft carriers, and airmail. It also took twenty years from the code for the Altair 8080 to the launch of Windows 95. So that puts a robot insect landing near your family picnic around 2034, give or take a few years.

Later that afternoon, Radhika returned to Reverend Billy. "He's right. We should save the bees. We should also spend more money on malaria, but we spend it on cancer." And I thought what a shame it was that the singers hadn't found a way to see Radhika as a person rather than a target or foil for their crusade about bees and Monsanto. It would have been a really interesting conversation.

A MONTH LATER I called Billy Talen, which is the Reverend's real name, just after he'd put his toddler down for a nap. On the phone he alternated between being a person and being his character, but he came across as essentially sincere. What he liked was bees, and, he'd gathered, so does everyone else: bees are a powerful emotional symbol, right up there with dogs and cats.

As he was trying to figure out how to make concrete the multiple abstractions of the financial, ecological, and political powers that influence the planet, the bee seemed like a good mascot. And then someone remembered that there were bee robots, and they Googled them, and they knew they had found their archvillain. "We identify a militarized honeybee. It's so immoral we don't even have to explain it." Later he put it another way: "It's apocalyptic. It's a complete capitulation. It's a horror film."

By the time we spoke, Reverend Billy had recognized that Monsanto was not funding the RoboBee, and that it had military implications. The horror movie concept was emotionally expedient—it got people to react—but it didn't really dig into the true complications of this technology. I found that frustrating. When he clowned on these topics, I thought Reverend Billy was forfeiting the ability to genuinely engage with the scientists and the rest of us. And the reason this bothered me was that when he talked about moral and political risks of swarming technology, he made a lot of sense: "Insect drone wars are like nukes: an impossible moral decision. And it's a decision that has to be vetted in the democratic process. We can't let this be decided by the military."

PART V

SEVENTEEN

DARWIN'S TERMITES

AUSTRALIA

AFTER A FEW YEARS of my termite odyssey I had seen them as bugs and as robots, as superorganisms, and also as models for a new economy, but I still hadn't gotten a sense of how these little insects make my world. I thought about termites all the time, I'd even imagined them inside myself, but I really didn't see how our lives on Earth—our actual evolutionary ecological fates—were intertwined. To understand that, I would have to go to Australia, the great continental experiment in the evolution of plants, insects, protists, and vertebrates.

And once I was there, Australia gave me an education in seeing what was not visible—white ants, hollow trees, and promises kept as well as broken.

But what actually got me to Australia was the desire to see something else. By 2011, I had a large collection of clippings about termites stuffed in a file cabinet. One folder contained information about the world's ur-termite, *Mastotermes darwiniensis*, the mastodon of termites, named after Darwin, the Australian city, which is named after the man. The top bug on my bucket list, *Mastotermes* are pretty exciting because they're the closest thing we've got to the first termites that evolved from cockroaches.

Nested inside the *Mastotermes* gut, though, is another amazing thing—a legendary protist named *Mixotricha paradoxa*: "the paradoxical being with mixed-up hairs." Under a low-power microscope, *M. paradoxa* looks like a grenade with a bad case of shag carpet, and it was discovered and named by a Jean L. Sutherland in 1933. Under interrogation, however, *M. paradoxa* turns out to be five entirely different creatures, with five separate genomes, collaborating as one, like a bunch of kids crowded into a donkey suit. The ultimate "composite animal," it has become a "poster protist" for big ideas about how such microbial complexity evolved, and it has acquired a subversive air, with a reputation as a natural cyborg. Although I knew it was absurd to have a crush on a protist, never mind one with five genomes, I wanted very badly to see it.

In early 2011, I got another email from Phil inviting me to go on a termite safari. By then Phil had moved to Australia, where he was co-leading a lab at the University of Queensland, in Brisbane, that was devoted to "ecogenomics," which is another name for metagenomics, but gives a sense of finding both ecological and economic solutions to problems. With a new team, Phil was doing work on climate change, coal mines, and lung disease, but he was still looking for termites. He wondered if I'd like to come gather termites in Darwin, Australia.

Of course I would! I bought a ticket and showed up in August.

Phil was waiting at the airport in Brisbane, where we both would board a new plane onward to Darwin. He looked exactly as disheveled as he had in the past, but he was alone. At the last minute the rest of his lab at the University of Queensland hadn't been able to come. So now I'd be the scientific enabler. Fortunately, we boarded the plane without losing anything.

As we flew north I looked down at central Australia's cryptic landscape, struggling to interpret it. Mostly it was a soft sandy brown, but sometimes small rectangles of bright color appeared, like semi-precious stones, with spiders of orange roads around them. Perhaps they were mines. The large wrinkles that looked like buried crocodiles covered with sand—perhaps a mountain range. Further north there was a speckled pattern of trees and brush, reminiscent of the Persian rug I'd seen while flying into Namibia. Here again, I could almost see a pattern, but then failed to sort it out in my brain. And finally, just before we landed, the tropical coast, with a fringe of sudden green wetlands edging into the paintbox-blue Timor Sea.

Darwin was damaged by Japanese bombers in World War II—memorialized in Bombing Road and an Air Raid Arcade—and ruined again by Cyclone Tracy on Christmas Day in 1974. Now home to a casino surrounded by eruptions of bright watered greenery, the city will be walloped by more cyclones and flooding as the climate changes. By 2070, more than three hundred days a year are expected to be over 95 degrees, up from eleven days currently.

We needed a four-wheel-drive vehicle to collect termites, and Phil's wife's uncle had offered to loan us one. Uncle David was a friendly pensioner with a yard crowded with exuberant greenery and shaded by an enormous mahogany tree. When he heard we were gathering termites, he volunteered that his mahogany tree was so infested with white ants that he was going to have to cut it down. White ants? "Mastos!" he exclaimed, as though announcing the name of the home team.

Only ten minutes into this country and we'd already found *Mastotermes darwiniensis*! We made our way through a pointy thicket of bird-of-paradise and other sharp and bright plants to the tree, peeled off a few strips of its scaly bark, and found hundreds of termites. They were bigger than any termites I'd ever seen, each the size and color of a grain of cooked white rice. Their heads were yellowish, their bodies translucent and full of guts. With their large size, reminiscent of their cockroach ancestors, they have been found in amber in Germany, Ethiopia, Brazil, the Dominican Republic, England, Mexico, and even Wyoming, where they appear to have gnawed on dinosaur bones. It is speculated that climate change in an earlier era caused them to go extinct everywhere except Australia.

"Hurry! They're going fast!" Uncle David shouted at us. Phil worried that he was "violating" the tree by pulling off the bark, but David brushed it off. "This tree's in big trouble, mate."

After my thirty-hour flight the mahogany tree was otherworldly and sinister. Peeling back its scaly bark revealed tentacle-like snarls of fibrous tendrils winding between bulging, shiny, tumor-like growths that were the shade of fuchsia you see on Hello Kitty. Unfamiliar birds chattered in the crown of the tree.

Uncle David was in a very good mood. "If I told people you came all the way from America and Brisbane for my Mastos, they'd never believe me."

Mastos, I quickly realized, are one of northern Australia's folk antiheroes, respected for their opposition to the greater national project of trying to extract a living from this place. People here like to brag about termites, who (they say) can be heard chewing in the walls at night. Everyone knows of a town where the lights went off because termites chewed through the power cables. There is supposedly a town named Termite but I can't find it on the map.

PHIL WAS NOT as carefree as he had been in California, and moments of jazz were rare. The science of metagenomics itself had grown up. The first termite gut metagenome study he'd worked on in 2007 had sequenced 63 million bases; typical sequences now were a thousand times larger. "The whole field is a hockey stick. Every diagram is like that," he said. The time he spent working with programmers to build new bioinformatics software to organize and interpret this data he described as a "race." But the result was that doing jazz no longer required the brainpower of him and Natalia—the field was rapidly becoming democratized. He had undergraduates working on complex tasks that would have been reserved for postdocs just a few years ago, and he foresaw a time when interpreting genomes would be "a parlor game." He was convincing on this topic mostly because he spent a lot of his spare time curating a microbial gene database called Greengenes, in the hopes of getting rid of that "shonkey" data.

Phil was still tremendously optimistic about science's ability to solve human problems, but his mood had changed. He still joked, and he still turned nearly everything upside down, but now he had responsibilities. Worried about providing for thirty-five people in the lab, he'd gone to a hypnotist to try to become more productive.

And maybe some of the optimism he'd had about using technology to address climate change had also worn away. The year before, Brisbane, where he and his family lived, had been hit with massive floods and evacuations. He described one of his friends, who'd been diagnosed with a liver tumor about the same time as the floods, feeling disoriented in his own waterlogged front yard, as if at sea.

Along one side of Australia, the Great Barrier Reef, the world's largest animal-built structure, was failing. The country has seventeen hundred plants and animals on the endangered species list. Those are only the creatures we count—which doesn't include many insects and certainly not microbes. "If the last blue whale choked to death on the last panda, it would be disastrous but not the end of the world," wrote the microbial ecologist Tom P. Curtis in 2006. "But if we accidentally poisoned the last two species of ammonia oxidizers, that would be another matter. It could be happening now and we wouldn't even know it." Phil's work in databases was not only about gathering knowledge; it was also trying to capture snapshots of an environment in transition.

But that was not all: a few months earlier he had reviewed a paper about factories that produced antibiotics in India. Unregulated, they were discharging antibiotics into the rivers so that the rivers had concentrations of the drugs normally found only in patients who were being treated. Bacteria in the rivers were evolving genes that were resistant to the antibiotics, and—as bacteria do—they were transferring those genes among themselves. Eventually they would leave the rivers and antibiotics would cease to work. There was a global experiment in health and the environment transpiring under our noses.

Phil had told me a story about his childhood. His parents had him and his siblings brought up into the clouds in an unpressurized Cessna when they had whooping cough. From there, hacking and shivering, they looked down on the patchwork of land underneath them. The first time he told me the story it reminded me of his ability to float above the world a little, as he did while doing jazz—his tendency to take a gently joking meta-view of data and the world around him. But when he told it again, I thought of those scared sick kids, and what it must have felt like to leave reason and comfort behind.

In the time since I'd driven down the Las Vegas strip with Phil, I'd changed my mind about hope. I no longer thought of it as a strategy. During the oil spill, I'd angrily watched destruction, despair, and politics intertwine and concluded that the gritty, emotional, annoyingly un-rational work of politics made the difference. We humans are like termites: when we are all running in the same direction, we can accomplish far more than the sum of our individual selves. More than that, when we are all running together, our world seems to make sense. Hope is emergent, rising up in the spaces between us. In 2010 technology had seemed like a way to jump over politics, but it didn't anymore. Technology was a different kind of politics: sure, it might solve problems, but the questions of who chose the problems and who chose the technologies were big. My interest in oil was waning and my interest in people was rising. How did people use politics to build real hope—a shared future—out of confusion?

THE NEXT MORNING we drove out of the city to a gauzy eucalyptus forest named Fly Creek, where we met an entomologist and two state government employees who were there to help us collect termites. Our plan was to get samples of as many kinds of termites as we could in the next two days and get them back to the lab's gene sequencers in Brisbane, where Phil and his team would analyze them and compare the inhabitants and genes of their guts. Some of this would go toward answering the Rosetta stone question of how termites got their gut inhabitants and proceeded to coevolve.

Hot breezes rustled the sparse dusty leaves above us, and the baked dirt below was pale pink, orange, and sometimes a terra-cotta that was almost lavender. I had to duck my head because the trees held biting ants with bright green globes on their butts that caught the sunlight like ornaments. "You can eat green ants but they taste like shit," said one of the guys. At my feet, there was a stick with droplets of red sap that glistened like fresh blood. "Bloodwood tree," said another guy. It was a bizarre, murderous landscape unlike anything I'd seen before.

The first mound was a small point, just a conical eruption of dirt about eight inches high that could be picked up like a chess piece. Inside was an elaborate maze made by *Microcerotermes*, a tiny termite whose soldiers have long rectangular heads and snapping mandibles. We collected some termites and put the cone back exactly as it had been.

We drove a little way and stopped to see a knee-high mound that looked like a melting scoop of pinkish-gray ice cream, or a malformed Buddha. We took turns knocking on its hard exterior. "You've got to hit it all around like a boiled egg," said one of the government men. After hitting it with a mattock, he lifted off its pinkish top so we could see a dense network of coral-like formations the color of red velvet cake.

Coptotermes live in trees in Central and South America, Africa, and Asia. Recent genetic work suggests that they arrived in Australia from Asia 13 million years ago, when the Arafura Sea was narrower. Probably, the trees they lived in fell in the sea and then—as if on insect Kon-Tikis—the bugs rafted over, landing on their new continent ready to continue their lifestyle. Australian trees, though, were small, and so the *Coptotermes* evolved to build mounds several times, in several different shapes, in different parts of the country.

We drove further, arriving at a field full of tall jagged teeth, in rows, maybe ten feet high and gray as battleships. They had a solemn, memorial feeling accentuated by their gravestone-like flatness. These were the so-called magnetic termite mounds, which are not magnetic themselves, though termites may use the Earth's magnetic field to orient their building. Like Scott's mounds in Namibia, these mounds are perfectly tuned to avoid the worst of the sun's heat. The Australian termite expert Peter Jacklyn studied these mounds by hitting them gently with a four-wheel-drive vehicle to nudge them into a new orientation. Once they were knocked out of alignment with the sun and shifted 20 degrees to the west, temperatures inside rose dramatically; shifted the other way, they fell.

These *Amitermes* are distant cousins of the termites Phil's lab collected in cow pies in Arizona, and though they build these massive mounds—which can be fifteen feet tall—in Northern Australia, to the south they just build simple cones. One of the men used a mattock to take off a chunk of the mound. It was as hard as concrete and shot through with tunnels like muffin batter.

I was copying down GPS coordinates in a lab book, and I was supposed to be recording all of this on video. When I watched the footage afterward, I discovered that the camera was always aiming in the wrong direction, picking up bits of conversation like "From here to Katherine it's all the same land." I looked it up later; Katherine was two hundred miles away. I realized I was worthless as a scientific assistant. I missed Anna, Phil's fearless lab tech in California.

The largest mounds came last: the cathedrals, more than twenty feet tall, had massive mud bulwarks with accessory piers, like the flutes on an organ in an old movie palace. Their surface was a patchwork of gray and gold sand that was, on closer inspection, a mix of fresh build layered over the cured walls. The *Nasutitermes triodiae* that built the mounds were truly tiny insects, about an eighth of an inch long. *Nasutitermes* mostly live in trees elsewhere, but genetic analysis suggests that they've floated over to Australia in logs at least three times in the last 20 million years. When the climate in Australia changed from moist and woody to dry and grassy, they evolved the ability to build mounds at least eight different times.

We ended the day with eight types of termites. Phil was delighted by the great variety and strange evolutionary quirks of the bugs. We spent hours in a government lab herding the live termites into containers so that Phil could carry them back to his sequencers. Phil put a termite under a microscope and shouted with delight when he saw it harbored an enormous mite. "That's like me carrying a dog on my back!" The government men graciously helped us sort—missing a rugby game for their trouble.

I realized that day that I had finally arrived in termite land—a place where the strangely plastic nature of termites was able to express itself fully through evolution. In Namibia and Cambridge I'd watched the roboticists search for the termites' algorithms; but here the termites themselves had changed their own algorithms—building little cones in one place and great towers in another. Did the local conditions alone offer enough signals to change the shape of the mounds? If, as Scott said, the mounds were a form of external memory, an inside-out brain, what were these termites thinking?

The bigger mystery, though, was something I didn't even see: 80 percent of the eucalyptus trees here in the north were actually hollow, eaten by termites. The person who told me this was Garry Cook, a scientist who studies fire and climate change at Australia's Commonwealth Scientific and Industrial Research Organization (CSIRO). "They're more or less hollow pipes. The inside of the tree is dead wood." He stopped and looked at me. "It's not like hollowing *you* out."

The idea that eight of every ten trees around us were hollow immediately struck me as hilarious. Never mind the ants, the bloodwood, and the termite mounds—if most of the trees were just straws, then I really couldn't trust my eyes. Once hollowed out, the trees burn differently. "The tops fall off and flames shoot out the top," Cook said, but the trees also produce different gases, and that's very important to Cook because about half the land in Northern Australia burns every year, and he studies how these fires interact with greenhouse gas production and prevention.

Eating standing trees is another way that Australian termites have changed to suit their surroundings. This part of Australia is profoundly shaped by fire. Aboriginal people have long used fire to manage the landscape and they arrived at least fifty thousand years ago. When Captain Cook arrived in the mid-1700s, he described the country as looking like parkland, without realizing that it was tended by its inhabitants according to rules and techniques passed down in song, stories, and artwork. Northern Australia's plants and trees have evolved to withstand and even take advantage of fire. And, it appeared, the termites had also evolved to take advantage of these relatively recent fires by eating the centers of standing eucalyptus trees. On an evolutionary scale, fifty thousand years isn't much time, but maybe termites— like dogs and cats—have long been involved in an unseen dance with humans in the landscape.

I FLEW BACK to Brisbane one evening around dusk. After the long expanse of darkness in the north, patterns of light appeared below me that looked like phosphorescent diatoms. One had a bright center with lights around it. Others were like spirochetes with a long line bent mysteriously in the middle. And then whole networks of lights and streets started appearing, growing denser, like filaments converging. It occurred to me that my worldview was getting increasingly microbial.

Because I wanted so badly to see *M. paradoxa*, the protist with five genomes, Phil introduced me to the microbiologist P. J. O'Donoghue. P.O.D., as he's called, was an excellent storyteller with a grand mustache. He said he'd decided to become "the godfather of protozoa in Australia" after surviving lung cancer. "We got sick of the American and European scientists gathering specimens and taking them away to write papers. It was a real bloody schemozzle for over a century."

As weird as the Australia I could see was, many of the animals here contained another equally weird Australia in their guts. P.O.D. described "a parallel universe" of protists in the poo of Australian horses and kangaroos. Termites make up 40 percent of Australia's faunal species, he said, so of course the lab started working on termites. "But they're small and colorless unless you bugger up the microscope."

Mixotricha paradoxa had been made famous by Lynn Margulis, the brilliant microbiologist who demonstrated that the mitochondria, the so-called powerhouses of our cells, were once free-living bacteria that made their way into another cell and then (with their ability to use oxygen) made themselves essential. The symbiotic relationship became closer until finally they merged. Similarly, free-floating blue-green algae once "jumped" into more-complex nucleated cells, to become the green chloroplasts that produce energy from sunlight within the single-celled ancestors of plants. This process of organisms combining their genes and fates is called "symbiogenesis," and Margulis said M. *paradoxa* was the "poster animal" for the concept, demonstrating a former time when individuals had this "composite nature."*

I really wanted to talk about M. *paradoxa*, but sitting here in Australia it was obvious that there were hundreds, if not thousands or millions, of similarly fascinating protists, and I had to drop my fevered questions about M. *paradoxa* or risk being the type of American boor who goes on and on about Crocodile Dundee.

P.O.D. introduced me to Linda Ly, a PhD student, who had figured out a way to use silver to make the protozoa visible. She'd already found thirty-two new protist chimeras—each with multiple genomes—in Australian termite guts. With a red bandana tied tightly over her hair, Linda had the blunt female variant of P.O.D.'s intensity. As a taxonomist, she'd been chasing dozens of previously undescribed giant protozoa with bacterial symbionts riding all over them. Like *Trichonympha*, some of these protists were a hundred times bigger than the bacteria in the termite's guts. Linda sketched their blobby outlines, examined the elements of their shaggy exteriors and the stuff inside them, and began the process of categorizing them. "One of my favorite species to look at is nice and big, it has a big tuft of 'hair,' and the nucleus is nestled in the side in a pocket. It seems to have a second cell wall. Wow. You can just look at it."

---

*Margulis looked at the microbial world with a very large lens, through a long span of time, and saw a dual explanation of how cells evolved to be complex while Earth's atmosphere evolved to include oxygen and cells evolved the ability to use that oxygen. She was also a great storyteller. "The speed, volume, and antiquity of bacterial gene-trading activities underlie the evolution of all the rest of life on Earth," she wrote with her son Dorion Sagan. "Humanity is a biospheric plume, an untested experiment, and may disappear in self-consumption."

Another protist was *Stephanonympha*: a single cell with multiple nuclei. Why? Linda answered: "Bugger if I know." Linda showed me a sketch of some of the thirty or forty protists she was working to identify. They seemed a bit human, and also worried and neurotic. They reminded me of old *New Yorker* cartoons by William Steig, with his scratchy Picasso-eyed "Lonely Ones." "They're so new you get a buzz from describing them," she said.

But when I looked at Linda's protists under the microscope, I saw just a bunch of rotten collapsed peppers, with indistinct edges. Concave and convex were indistinguishable. The squishy obscurity of it all made me anxious, as if I were one of the Lonely Ones myself. Even if I met the famous *M. paradoxa* on the street, I wouldn't know it!

Later I called Patrick Keeling, the Vancouver protist expert. He told me the ancestor of the nutty creatures that Linda found was probably a small, relatively unimpressive protist with four flagella. The peculiar environmental conditions of the termite gut supported the evolution of their structure, behavior, and symbiotic relationships, many times over, in both similar and strange ways. How did the little flagellate make itself a hundred times bigger, enabling it to eat really big wood chips? The answer seems to be that it repeated its structural elements along a line of symmetry, as if bolting one IKEA bookshelf to the next until it had something the size of a library. In some cases, as it duplicated these elements, it may also have duplicated its nucleus. (This is, at the moment, the best answer for why some protists have multiple nuclei.)

And the relationship between the protists and their spirochetes is surprising. Patrick studied one protist that looked like longish ridged starfruit with a dense curtain of symbiotic bacteria hanging off the outside. When he dosed it with antibiotics, the bacteria fell away as expected, but without them the protist itself lost its ridges and length; it literally went pear-shaped without its symbionts!

The one thing that these termite protists aren't, Patrick came to suspect, is poster protists for symbiogenesis. The genomes of these protists were not on the verge of combining for good, but—in Patrick's view—many had genes that were "an increasingly strange mutational circus," and he suspected they'd soon go extinct in a "blaze of genetic weirdness." The reason for this is that the longer a symbiont stays with a protist, the more funky the symbiont's genome gets. With food and reproduction assured by its host, all kinds of messy DNA code makes its way into the symbiont's genome and—as long as it's not harmful—just stays.

These odd marriages of protist and bacteria, then, are probably not snapshots from a former time when symbiogenesis was common, but very peculiar products of the futuristic junkyard of Australian termites' guts.

Patrick wondered how scientists had ended up using symbiogenesis as an explanation for relationships they couldn't explain. He began studying the history of research on cells. Old-school biologists had done a lot of structural work on how these protists evolved. But when symbiogenesis became the accepted paradigm for the evolution of mitochondria, and piles of genomic data arrived right after, scientists stopped looking at the older papers. Patrick turned his microscope on the scientists themselves, writing that it's "almost as though we are ourselves experimental subjects but never get the diagnosis needed to be more self-aware of the reflexive nature of scientific theories."

I never did see *M. paradoxa*, but in learning that the poster protist was not what it seemed, I saw something bigger about the nature of science's search for truths. Like stigmergy and the superorganism, the symbiogenic protist was the kind of idea that could inspire scientists for generations, even if it did not turn out to be, strictly speaking, true.

EIGHTEEN

THE SOUL
OF THE SOIL

I HAD BUILT AN EXTRA WEEK into my trip to Australia, figuring I would take a "real" vacation. But I didn't know how to do that: I was more comfortable thinking about termites. In retrospect, my inability to relax seems a little pathetic. On the other hand, it was exactly through this inability to stop obsessing that I found the story.

In a hotel room above a drive-through liquor store I read the newspaper and watched TV. In early August, the production company for *Mad Max: Fury Road* announced that climate change made it impossible for them to shoot in the country, so they were moving to Namibia. Quintessentially Australian, the Mad Max films are about people who've exhausted their resources and now run madly around in crazy vehicles through a postapocalyptic landscape searching for fuel. But when the production company went to film at Broken Hill, New South

Wales, a notoriously dry spot, they discovered it had gotten so much rain that its fields were waving with wildflowers and the dry lake suddenly had pelicans. Ironically, Namibia's fairly healthy systems of sand and lichen crust made for a more cinematic apocalypse than the truly disastrous wildflowers of New South Wales.

I continued reading about termites. There was a paper about a remote bauxite mine where termites had rehabilitated more than eleven square miles of land over the course of twenty-six years. Studies never encompass twenty-six years, and rarely so much space. I wondered why this one had carried on for so long.

Another Australian study reported that a farmer had increased the amount of wheat he grew in his dry fields by 36 percent when termites and ants were allowed to do their thing in the dirt. The authors speculated that termites move the soil around, which makes it hold more water in dry areas, and that their gut microbes also increase the amount of nitrogen available for the wheat to use. One of the termite's selling points was that farmers didn't have to spread expensive nitrogen fertilizers on their fields.

Termites could be the ecosystem engineers humans need. Even if we eventually use synthetic biology to grow hamburger in tanks, we will need to figure out how to grow more food and restore land that's been abused by mining and pollution—while trying to maintain forests and savannas going through the turmoil of climate change. In 2050, as the population of the planet peaks, we'll need 60 percent more food than we currently grow to feed increasingly affluent people. And if synthetic biologists do manage to make grassoline, we'll need to increase the amount of green stuff we grow per acre between two and three times. At the same time, climate change will be reducing the fertility of the farmlands belonging to the poorest people. Whether we can resolve these challenges with decency is a test of who we are.

I tried to contact an author of the wheat study, but she'd left CSIRO and didn't reply. I also wrote to an author of the mine study, a former CSIRO scientist named Alister Spain. He replied that he had retired to Magnetic Island, and if I visited he'd be happy to talk. I checked my guidebook: Magnetic Island was a tourist destination in the Great Barrier Reef Marine Park on Australia's east coast. It had gotten its previous name—Magnetical—from Captain Cook in 1770.

I told Phil where I was going and he suggested that I stop along the way, in Townsville, to meet up with a young scientist named Rochelle Soo, try to find some termites, and talk to an exterminator he had heard of. I went to Townsville and met Rochelle; we failed to find many termites but we did find the exterminator. He was middle-aged and wore shiny metallic slacks, a tight blue-striped shirt unbuttoned part of the way, and a large number of prominent rings on one hand. He enjoyed having an audience and had seen with his own eyes all the legendary feats of *Mastotermes*. They ate not one but two tractor tires. They'd chewed through a two-story house in just a day. They reduced a whole house to ruins during the owner's two-week shift at a remote mine. I supposed there was truth in there. "They've made me very rich," he said. His plan was to buy a semitruck with a sleeping cabin to pull a four-wheeler with a spray setup on it to the Outback, where he'd go from farm to farm giving relief from the termites.

I wrote "Termite Truckers" in my notebook to remind myself of a quick movie idea I'd had. A manly termite killer, ideally Mel Gibson, drives across the outback in a semi and encounters a mad scientist whose brain has been invaded by termites, causing her to molt and begin sensing gas gradients beyond her body. Tilda Swinton? There will have to be a trophallaxis scene. The two characters have different ideas of how to save the world: one wants to kill the termites and the other wishes to nurture them with her sweat so the termites do the work. They are racing each other in giant trucks. "This doesn't turn into a love story," I wrote.

I took the ferry to Magnetic Island, checked into a cheap room near the beach, and borrowed a snorkel. The rules of the dying Great Reef upend old common sense. Now you can't wear sunscreen near the reef because fourteen thousand tons of sunscreen lotion every year are accelerating the bleaching of the world's coral. I did the dead man's float in a small lagoon, letting the coral make psychedelic patterns in front of my eyes. It looked like the inside of a *Coptotermes* mound, an alternate universe straight out of Ernst Haeckel's *Art Forms of Nature*, but I had no idea whether it was healthy or even dead. Floating and snuffling through the snorkel, I sloshed around—a piece of flotsam growing pinker by the minute.

When you put a termite down in a new environment, it will zigzag around, sniffing the dirt. I think of this time in Australia as if I were an aimless foraging termite. I didn't focus much. I dabbled in one thing or another, sloshing about, seeing termites everywhere but not really knowing what I was looking for. As with the rest of my study of the subject, it was not efficient, but it was in its way exhaustive.

The next day I met Alister Spain in his wooden house overlooking a tree-filled gorge, so that we seemed to be flying as he served tea. Alister had a narrow face, gently hooded eyes, and a calm, curious demeanor. He had spent much of his career in far harsher places, studying the impact of mining bauxite, coal, and uranium on soil. "Part of the mentality of Australia is a mining mentality," he explained.

I asked him why this termite restoration program had gone on for twenty-six years. The real story, he said, was longer and more complicated. The mine was on land belonging to an indigenous community that had petitioned the government to stop the mine. When they lost the court case, they asked the mining company to return the land in its original forested condition. In the 1970s, a young German named Dieter Hinz, a horticulturalist whom Alister called "an enormously capable observer of natural history," came up with a scheme to bring back the forest. It worked, and proceeded almost without change to 2005. Because of the political significance of the rehabilitation work, the mine's bosses gave Dieter resources. After about sixteen years, the mining company wanted verification of their methods, and they asked CSIRO to study the project. Alister was part of that team and he took part in later studies of the project. To learn more, he said, I'd need to talk to Dieter.

A few days later I took a bus to the Queensland town of Toowoomba. In his seventies, Dieter was compact and muscular, with a neat, square beard and a blunt cap of hair. We went to a slick, heavily air-conditioned café where the Europop was fast and loud. He seemed out of place, arranging his hands carefully amid the silverware. I liked him immediately: his exterior neatness seemed to speak to a deeper interior order.

Anyway, he was also crazy about termites. Over the next few hours he told me a strange tale about bugs, his initiation into an Aboriginal clan, and how forest destroyed by bauxite mining came to be restored, complete with enough stringybark eucalyptus trees to keep the clan's ancestors happy.

But first, some backstory: In the 1950s, a huge amount of bauxite, the ore used to make aluminum, was found just under the topsoil of the Gove Peninsula, a remote area in the far northwest of Australia where people who spoke a group of languages called Yolngu had lived for thousands of years. The government removed 115 square miles of land from the Yolngu reserve and gave it to the Swiss mining company Nabalco without consulting the residents.

In 1963 the Yolngu elders wrote a letter on two pieces of bark asking for Australia's parliament to do an inquiry. In the center of the bark is the letter, typewritten in English and Gumatj, a Yolngu language, and around the edges are drawings of fish, turtles, snakes, and other animals. The bark petition is now a famous artifact of Australian history, but at the time it was a radical assertion of sovereignty and a display of local art that changed the way the country talked about the mine. Parliament declined to do an inquiry and the Yolngu took their case to the Australian Supreme Court in 1968.

In court, Yolngu leaders and anthropologists were asked to testify about Yolngu beliefs regarding land. The anthropologist Nancy M. Williams, who lived for many years in the area, wrote a book about the trial. It's impossible, she wrote, to separate land tenure and political structures from religion and its expression. "It seems to me that Yolngu perceive time as circular, so that from any particular time, what is past may be future and what is future may be past. . . . Yolngu souls progress through states of being: inchoate, but sharing the essence of their clan's central spirit-being ('totem'). They exist at focal sites on clan land; they enter a foetus to animate it; they depart at death . . . to return then to a site on clan land and be available to animate another Yolngu foetus."

The judge listened to complicated testimony on the subject and leapt to his own patronizing interpretation: the Yolngu did not so much own the land as the land owned them. Though he acknowledged their connection to the land, the judge said it didn't apply to the legal system of Australia. And so mining went ahead at Gove.

In the late 1960s, Dieter, who had left Germany in search of adventure after getting training in horticulture, was wandering around in the deserts of southern Australia looking for uranium, a dog by his side. Hired by Nabalco to restore the mined land in 1969, he traveled to Gove. When he arrived, the mine wasn't ready and he was told he could be the caterer. Six months later he started a plant nursery, employing local men and, gradually, local women. "The men didn't like the women working for someone who they didn't know. So I had to work up the relationship." It was the women, though, who understood the seeds and the landscape. "They knew the relation with the plants—medical, totems, or food. It was a collaboration. I learned what I wanted to know and why."

Bauxite miners don't dig holes—they scrape, as though removing the center from a layer cake. The bauxite ore itself is a horizontal pile of pea-shaped ore chunks lying in layers between three and thirty feet thick. To get to it, miners fell the trees, bulldoze them, and burn them before scraping off the first foot or so of topsoil and storing it. Then they remove the next eight feet of subsoil and store that. Then they load that layer of reddish bauxite peas into trucks and drive them to a nearby refinery. Finally, their bulldozer blades hit the mine floor, a hard layer that may be ironstone or hard clay. They rough that up, then replace the subsoil and finally the topsoil. And there: the land is back—several yards lower, with no plants. This was what Dieter had to work with.

When the mine began operation, he had to change the way the bulldozer drivers worked on the region's fragile soils. "Holy moly. Pushing the soil up and dumping it somewhere, it's just not on. The soil is alive! The termites are in there and the fungi are in there and the bacteria are in there."

The wet season was a critical time for the soil, seeds, and fungi, so Dieter got the engineers to move soil mostly during the dry. And he also got them to leave the tree roots in the soil so that the fungi stayed in place. "As soon as the trees shoot out again, the mycorrhizae says I'm here. I'm fine. I'm waiting for this." Dieter liked to walk through the fields collecting soil. At night he'd look through his microscope at all the things living in the dirt. "You have a different world there."

The mining company initially expected to replant the mined land with grass for cattle grazing. But the Yolngu said they didn't want cattle, or even forests of nonnative trees like mahogany. In 1972, at a meeting with the Swiss mine manager, the elders said they wanted the land back in the original condition, with eucalyptus trees rustling overhead. Making this happen became Dieter's task.

One reason the community wanted the area regrown is that some specific plants and animals are totems to the families or clans. "They explained to me that, Dieter, you have to bring back the stringybark tree—gadayka—and you have to work fast." When Dieter protested that he also had to answer to the mining company, an elder invited him to join his clan. "He said, 'Now you belong to our clan, which means that you have to bring that tree, the *Eucalyptus tetrodonta*, back. It's the totem of the clan, so now it's your duty.' Well. I felt odd saying I belonged to the clan, but then there is a demand that I make sure that those trees are definitely coming back." And once back, they had to stay healthy and grow. Dieter laughed at the hugeness of the request with a breathless cackle, spreading his hands to show his happy bewilderment.

Dieter's plans were unconventional by the standards of the early 1970s but are close to state-of-the-art now. He hired local people to identify native plants and collect their seeds. He controlled the burning and storage of the topsoil so that it never had to wait more than a year before it was used again, preserving seeds and spores in the soil. And he avoided applying nitrogen fertilizer to the ground. Covering the ground with too much nitrogen would destroy the symbiotic root fungi necessary to grow trees. Instead, he applied some phosphate, which is low in those soils.

This unconventional approach made the mining company anxious, so Dieter offered to quit if he hadn't had some success after two years. But the rehabilitated land was quickly obvious: the grasses grew back. Later he hired large crews of local workers, who stored hundreds of pounds of seeds, not only for the next year but the year after. He slowly got to know the landscape and the people together, by asking questions. At first he didn't get answers. "Sometimes you're off the subject because it's a woman's business or a man's business and they don't want to get into all of it. So you let it go and six months or a year later it comes up again and bingo!"

Over the years twelve kinds of termites came to the site, like a series of twelve choreographed bucket brigades. When there was only grass, two kinds of *Amitermes* came. And as brush grew up, other termites capable of digesting wood tromped in, and the grass-eating *Amitermes* moved on. By the time there was forest and the canopy closed in, the mix of termites was different.

Eight years into the project the bees came back, and that made workers happy because they could look for honey when they were working in the fields. The mined land started becoming part of the continuum of present, past, and future. Birds returned. And then a certain type of tree started flowering. "Once an old fellow told me, 'Look at that plant over there. It's a wild yam. We don't eat it anymore, but in the early days we did. Now the spirits of our ancestors will be able to come back here.'" Earthworms showed up in the sixteenth year.

Eventually there was a forest of eucalyptus trees, a lively midstory, and many native plants, including twenty edible ones. Within the soil, there were different types of roots and fungi associated with them. And above there were birds, stripy echidnas, wallabies, and lizards. And there were a lot of termites—eleven hundred mounds per every two-and-a-half-acre plot—revealing the vibrant social scene underground.

He had a close friend, Gambali, who named his son after Dieter. Now Dieter hopes to name a termite after Gambali. "He passed away. I spent about ten years working with him. It was very much so, yes . . ." He was silent, looking at his hands, and the harsh racket of the restaurant felt suddenly unbearable. "A deep relationship," he finished.

The scientists who came to examine the site had a lot of pre-conceived notions, and Dieter didn't tell them everything he knew. "Everybody was an expert. They were keen on transects. And then they'd do a count. I said that's ridiculous! You can't put a finger on an ecosystem!"

Alister was a different sort of scientist, though. He not only re-spected Dieter's knowledge, he had a deep interest in soil ecology and the ways that earthworms, termites, ants, roots, and fungi created structures and fertility in the ground. Twenty-five years later they were still writing papers together.

Eventually, Dieter got asked to do a rehabilitation project at an old bauxite mine in Sierra Leone. He stayed at Gove part of the year but went back and forth to Africa. The mine project was going well, and he even managed to smuggle some termite specimens out in little Un-derberg aperitif bottles so he could get them properly identified. But then came a civil war, the mine closed, and the families who worked there fled.

Around 2005, the Gove mine's new owners decided to abandon Dieter's way of managing the land to rehabilitate it more cheaply. Dieter left, retiring to his ranch near Toowoomba. "It sort of takes you out of the picture." His wife had always been just politely interested in inver-tebrates, but now she was ill and he was caring for her. These days, he was organizing his collections of rocks and fossils and memories. He had a lot of unfinished work to do recording his observations of the termites and their lives in the soil. "I've got a life story to write about that!" Dieter said with a laugh.

NINETEEN

THE MATH OF
FAIRY CIRCLES

NEW JERSEY

AFTER I CAME BACK from Australia, I wondered how termites rehabilitated the land that Dieter worked on. I wondered if there was more to them than the fact that they fertilized the soil and recycled the dead grasses. There seemed to be a gap between bugs dropping a few extra nitrogen molecules in their poo and the creation of a whole forest. What were they *doing* down there? I started going through my files looking for people working on landscapes.

This led me to the work of a mathematician named Corina Tarnita and an ecologist named Rob Pringle. When I contacted her, Corina had just moved to Princeton from Harvard and, with Rob, had set about using mathematical modeling to figure out what termites were doing in dry landscapes in Kenya. As it happened, I had interviewed Rob back in 2010, when he and a team published a paper on the role of termites in the African savanna ecosystems that are home to elephants and giraffes.

I took the train to Princeton to meet them in early 2014. By that time I'd pretty much given up on Carnot ever showing up or on global meeting local in front of my eyes. I had little hope that termite science would save the world. But through mathematical models, Corina and Rob and their team eventually delivered a version of those things. And they might have solved the mystery of the fairy circles, too.

Corina was teaching when I arrived at Princeton, which is a mixture of old buildings with weather-rounded bricks and cautiously futuristic buildings made of glass and new bricks, so I went to find Rob. Back when I talked with him on the phone, Rob had been using lizards to understand ecosystems. He'd map off a section of land, cut it into a grid, and go in and count the lizards as an index species. If there were lizards, there were bugs, and if there were bugs there were plants, and if there were plants there was some water. As it happened, places with enough plants to attract lots of lizards were also favorites of elephants. And elephants were, in a way, the point of these studies: everyone wants more elephants.

In the relatively dry savanna in Kenya, Rob had been finding between three hundred and eleven hundred geckos in his plots, but there were two places where his lizard counts went through the roof: in abandoned cattle corrals filled with dung, and on termite mounds. The dung was obvious—it would fertilize lush plants that would attract insects—but the relationship between geckos and termite mounds was not.

Macrotermes in that part of Kenya build most of their mound underground, so they look less like the fingers I saw in Namibia than like land with a case of chicken pox, with each bump of a mound situated twenty to forty yards from other bumps on all sides. The closer he was to the center of the mound, the more geckos Rob found. So then he looked at the bunchgrass and the acacia trees. A similar pattern. It was as though the termites had organized the entire landscape from below into a large checkerboard of fertility. "The termites are unwittingly pulling the strings without coming out of the ground," he had said when I called him in 2010.

Some part of termites' influence had to do with nutrients: a team of scientists found that the soils in the mounds were much richer in nitrogen and phosphorus than those off the mounds, and as a result the trees and grasses were not only more abundant there, but also had more nitrogen in their leaves, making them more nutritious—and possibly even more delicious—to everyone eating them. The termites also moved sand particles, so water behaved differently on the mounds. I asked Rob whether the termites were "farming" the land to get more grass to eat. He said that while it was clear they were caring for their belowground fungus, the mechanics of what was happening aboveground were unclear. It could be a series of feedback loops that resulted in more for everyone. Part of their impact was that the soil around the mounds held water differently, but how exactly that happened wasn't clear. "Termites are really important at regulating water flow. They're a black box."

When I first heard Rob talk about the black box, I understood it as a metaphor rather than an engineering concept. Now, as I looked over my old notes, I wondered what he had meant by it: Was he really looking to engineering to answer an ecological question? Or did he also mean it metaphorically?

What really bugged Rob about termites was the pattern they created on the land. It was as though the termites created a lattice that turned an otherwise monotonous plain into a series of hotspots. Something was going on with the way the space was organized that made the whole system more productive. And with the advent of remote imaging technologies like LIDAR (Light Detection and Ranging), which uses lasers to create images of the ground, these patterns were showing up all over. The amount he did not know had bothered Rob: "I can't not notice these patterns whenever I get in a small plane or look at Google Earth." I knew what he meant, because I'd seen similar rug-like patterns in Namibia and Australia. When he was at Harvard, he had shared his irritating problem with Corina.

I knew that Rob had a fine sense of the absurd, because I'd run into a photo of him online wearing a black suit while wrestling with a tape measure near an electric fence. When I found him in his office, he was wearing jeans and cowboy boots amid Princeton's old brick.

Corina arrived from class in high boots and a striking dress, poised and glamorous in the bare junior faculty offices. She also has a profoundly cerebral air: she takes things in, refracts them through a mathematical prism, and sees them in an entirely new way.

Corina grew up on a farm in Romania, was fascinated by math early on, and won multiple prizes before heading off to get an undergraduate degree at Harvard. She started her master's program there working on something called high-dimensional geometry, but switched midway to mathematical biology, where the questions were more real and messier.

In 2010 she published the big paper with the biological mathematician Martin Nowak and the entomologist E. O. Wilson questioning Hamilton's theory of inclusive fitness, which had been based on J.B.S. Haldane's bar bet on how many relatives he would risk drowning for. Corina had spent a year doing the math and found that the fact of being related alone wasn't what made cooperation successful. When a queen could produce daughter ants that would stay and raise her brood, more of her babies survived: cooperation produces more kin.

In 2013 she went with Rob to the Mpala Research Centre in Kenya. There, instead of modeling competition and cooperation on her computer, she could actually play with termites from different nests to see how they fought when they were put together. At first all of the flat grassy land they were studying looked the same, and she had a hard time picking out where the termite mounds were. But as she got used to finding that pattern, she started getting a funny feeling.

"A million questions are triggered in the field. I could see that there were more patterns than just those of the mounds. I sensed a pattern but I could never quite pick it up." One afternoon, after they'd been working on their National Science Foundation grant proposal for three weeks, they went for a walk. They went past a field that had been burned, so the vegetation was just stumps of grass, not the waving tops. Corina thought she saw something and asked to stand on the roof of a Range Rover. And then she saw it, among the burned stumps: two separate patterns interacting with each other. First there was the polka-dotted pattern of the mounds, and then she thought she saw a leopard-spot pattern in the vegetation between the mounds.

As starfish patterns are to roboticists who work on self-assembly, so leopard spots beckon to biological mathematicians—a natural shape with a theory behind it. The leopard pattern resembled a Turing pattern, a theoretical construct that was first proposed by the British mathematician Alan Turing in 1952 and was subsequently demonstrated in some natural systems. If you've seen leopards, zebrafish, zebras-who-are-not-fish, seashells, and chameleons, you've seen these patterns, which are sometimes called "reaction-diffusion" or "scale-dependent" patterns. They are essential components of the world's organization, influencing everything from how slime molds form in sink drains to the way rabbit brains perceive smells.

When Corina told Rob she saw leopard spots, he was skeptical and said it was just clumpy bunchgrass. But she insisted, and so he took some photos. Later they sent grad students out to take more photos with a camera on a thirty-three-foot pole. And when they examined these photos, it was clear that another pattern *was* in operation on the ground. Rob quickly realized his error—he had simply known too much about the plants to really see past what he knew. "It's awesome to be in the field with Corina. It's not surprising, but she didn't have the same ideas about root competition that I did."

Corina, for her part, was in heaven. "Before Rob brought me to Africa I was a theoretical biologist. I now almost don't want to work on systems that I can't see or manipulate. That's a big change."

But that moment was only the beginning of the work she and a team had to do to build a model, verify a hypothesis, and use it to predict how these interacting pattern mechanisms would look in nature. "What comes before the model is an intuition of what the rules for the patterns could be," she told me later. "I put that together into a skeleton and then I add in a lot of detail about how the termites and the plants actually function. It's like detective work."

Working with Juan Bonachela, a Scotland-based theoretical biologist trained in statistical physics, and Efrat Sheffer, a Jerusalem-based biologist who studies the relationship between individual plants and their ecosystems, Corina began to build a model, starting with a proxy for how termite mounds organize the landscape: a simple lattice of hexagons. Termites leave the mound to forage in an ever-widening circle, but over many decades, as a landscape gets filled with termite mounds, the foraging zone of each mound bumps up against others. When all of the mounds contain roughly the same number of termites, they end up spacing themselves across the landscape evenly. Where the radius of foraging termites from one mound hits the radius from the next, they form an edge. It's not a perfect border, and it's not visible aboveground, but it's there anyway, perhaps caused by the ferocious fighting their postdoc Jessica Castillo-Vardaro saw when she put termites from two mounds together, or maybe caused by termites avoiding other termites who don't smell like their kin. If the mounds are distributed evenly across the landscape, most mounds will have six neighbors. So in the end the mounds look like a patchwork of hexagons, maximizing the distance between every mound. This made sense: many other creatures have self-organized hexagonal territories, including wolves, Alaskan sandpipers, and even some kinds of fish.

Next the team built a model for the pattern in the grass. The basic concept of Turing models is that there are two different feedback mechanisms. Within a short distance growth is encouraged (activated) and over a long distance it's discouraged (inhibited). For example, plants that are close to each other may help each other by more efficiently absorbing water from rainfall, creating a little tuft, but over distances the tufts begin to compete with each other for water, suppressing growth and leaving bare dirt. In a model, if you give the activator and the inhibitor each four different parameters (how fast they diffuse, how strong they are, and so on) and build a model where they interact, over time they will form a characteristic set of patterns. Those with a strong activator, say, will form large spots, while those with a strong inhibitor will form small polka dots. Play with the parameters for diffusion at the same time, though, and you'll get patterns that look more like a tortoise shell, a bunch of donuts, gaps, stripes, or a labyrinth. These patterns look very much like things in real life, including mussel beds, coral reefs, and fungi. So biologists have speculated that the activators for such scale-dependent feedback could be as diverse as water, hormones, or organisms helping each other, while the inhibitors could be drought, hormones, or predators.

Using a mixture of activators and inhibitors appropriate for the grasses in Kenya, Corina and Juan built a mathematical model that created scale-dependent feedback patterns. Playing with the parameters on the grasses alone, they could limit the water and turn the tufts from big tufts to a labyrinth and finally to desert, which looked a lot like the photos. "We said, 'Okay, there's plenty of reaction diffusion,'" she said of the grass patterns. "Can we couple that with the termite patterns?"

She combined the self-organized termite mounds and the scale-dependent feedback model for the grass. Now, instead of an even pattern of polka-dotted mounds in the middle of hexagons, or a leopard pattern of grasses, the leopard pattern arranged itself over the hexagons: lush over the resource-rich mound in the middle and sparse at the edges of the mound. Corina printed out images of the two models interacting, with different amounts of rainfall, and showed them to Rob.

In very general terms, the images looked like African patterned cloth: regularly spaced large dots surrounded by halos with a kind of calico background pattern. The dots were the mounds and the calico was the leopard pattern in the grass. When Corina and Rob compared the models with satellite images of termite landscapes in Africa, they found that they looked very similar. They could even zoom in on the calico in the model to see the shapes of the bunchgrass, and they found it resembled the shapes in the photos they had taken. These patterns had previously been hidden in plain sight. "It was the convergence of the model predictions and the data that made me believe," said Rob.

For Corina, the thrill was finding that the two different patterns, at multiple scales, were interacting and influencing each other. The local was connecting to the global and it was even showing up on the satellite maps. "I'm happiest when models can be tested and we find so much agreement," she said with obvious delight.

For me, the math of Corina's team explained the unsettling feeling I had looking out of planes in Namibia and Australia. That sensation that I could almost see the pattern of a Persian rug in the landscape had been correct. And now that I could finally really see the design in her modeled images, I thought about how Corina's intuition had combined with the powerful math of the models to reveal something new. These models were a form of the new "alphabet" Eugène Marais had said we'd need to see termites as they really are.

But running the two models had provided another insight, with much bigger implications. In fiddling with the rainfall on the mounds, Corina discovered that when grass was associated with a termite mound, it could survive on very little water, much less than expected. In the simplest terms, termite mounds made the landscape much more drought resistant.

This observation had a practical benefit. Biologists had used patterns of labyrinths and spots to predict that some dry landscapes grew patchy just before catastrophically shifting to become deserts, which is a great fear in both Africa and Australia. Those theoretical models, from the mid-2000s, predicted that when these dry land systems crashed, they wouldn't gradually dry up but would instead progress from a labyrinth pattern of grass to spots, and then basically fall off a cliff (called a "critical transition") to become desert. Recovery would be very difficult, if not impossible.

But when Corina adjusted the rainfall in the model to produce the labyrinth of plants that might precede a crash, she found that when a landscape had termite mounds, the crash occurred very slowly—it was not a cliff but a staircase. What this meant was that places with termite mounds were much less likely to become desert, and if they did, they were likely to recover when rains reappeared. As long as the termites remained, grasses would sprout first on the mound and then in distinctive patterns. Termites, then, appeared to increase the robustness of the whole place, in addition to providing homes for the geckos and food for the elephants. And with dry lands making up about 40 percent of the world, and climate change redistributing rainfall, termites might actually be saving the planet. For real. For once.

The model was nice, but models are a pseudo-world. The team's next step was to further test the model's predictions with experiments in the Kenyan fields. By giving some mounds and their surroundings extra water while preventing others from getting rain, Rob and Corina and their team hoped to see whether the grass patterns would change as their model predicted. In order to do that, they needed to figure out how to block rain on some plots while increasing it on others. A fellow ecologist at Princeton was doing something similar in the Pine Barrens, a large forest in otherwise suburban New Jersey, and they said I could tag along when they went to see his structures.

It was a cold day in November, and the Pine Barrens lived up to their name: miles of tall dark pines, scraggly in late fall with relatively clear ground under them. In the woods, the researcher had built careful little shelters out of two-by-fours, using Home Depot hardware, to prevent rain from falling on some plots, while sprinkling others with extra water. Under the pines it was dark, and even chillier. I had worn only a fleece and I tried to conserve heat by hunching.

Rob thought that the elephants weren't going to like the little houses guarding an oasis of scrumptious-looking grass in a dry savanna. "I think elephants are a generalized stochastic hazard, but they're going to be very attracted to the water." He doubted that they could build something strong enough to keep them out. The elephants were wily, too. Even electric fencing would have drawbacks. "If we put fence that's two meters high, the elephants will play with it and mess around, but the giraffes will run right into it because they're not paying attention." It was funny to stand shivering under these dark pines in New Jersey, talking about the fields of inattentive giraffes.

On the way back to Princeton they mentioned they'd had informal conversations with colleagues who questioned whether the patterning was the result of termites. Rob felt that some of the skepticism came from scientists unfamiliar with self-organizing systems, who might think that such large-scale pattern organization implied a "mastermind." And some ecologists assume that if competition among termite colonies is strong enough to drive this elaborate patterning, then it would be likely to push the resource base toward collapse. (A sort of Mad Max theory.) The idea that termites could be competing so strongly that they create patterns while making the ecosystem less likely to collapse? "It's a hard hump to get over."

Back in her office, Corina explained that her next plan was to work with the team to build a far more detailed model of both termites and grass with which to go to the Olympics of the patterning debate: fairy circles. Fairy circles are mysterious, roundish areas of bare dirt found in Northern Namibia and also in Australia, generally surrounded by grass. In aerial photos they look like regularly spaced pinkish elephant footprints of dry dirt, ranging in size from about nine feet to ninety-eight feet across. They've been the subject of fierce debate between scientists who say they're made by termites and others who say they're made by grass patterning. Though these patterns have been the subject of scholarship since the 1970s, interest in them spiked between 2012 and 2016, when a small spate of papers attributing the circles to one thing or the other appeared in journals. Corina felt that with a more complete model she could show that the fairy circles were the result of the termites' self-organization and the grasses' scale-dependent feedback combined.

Building the model was difficult, though. "The model forces you to have a rule for everything. You can't have any blank spaces," she said when we spoke early in 2015. "It forces you to consider what you might not consider otherwise." The termites were toiling away in their black box underground, unknown. She was deep in termite literature, and communicating daily with Juan in Scotland as the team built the computational aspects of the model. She said it was the most complex model she'd ever worked on, and she was facing inconsistencies in the thinking about scale-dependent feedback: the idea that plants benefited from being close to one another made sense, but did competition really suppress growth on a large scale? Another question was how termites concentrated nutrients in the space—of course they brought grass back to the mound—but they also processed some in their guts, such as bio-available nitrogen. It was a big puzzle.

"For me it's not the fairy circles," she said. What she wanted was to understand how multiple pattern mechanisms could interact at multiple scales. "I think it's fascinating that these little organisms, which are part of messy and complex ecosystems, can produce regular patterns."

In 2017 the team, which also included two grad students, published a paper that modeled how burrowing animals like termites, ants, and rodents might interact with grasses to create vast patterns and structures around the planet. Adding the termite lands of Africa, Asia, and Australia to similar earthmound-field landforms like Brazil's murundus, the mima mounds of the Pacific Northwest, and the heuweltjies of South Africa suggested that many tens of thousands of square miles may have been re-ordered from below. No mastermind could possibly have pulled this off: only trillions of mini-minds could possibly have taken on a task this big.

Now that I could see this relationship between the tiny diggers and the great scope of land from the air, I felt sympathy for the early scientists who looked into termite mounds and saw only human organization and proof of the rights of kings. By looking inward, they missed seeing the earthly equivalent of the celestial spheres.

In the mound, it is possible to see the whole order of the terrestrial sphere, or in more modern language, the progress from local to global. First there is the teeming world of the termite's gut, processing grass; then the world of the termites, digging and grooming in their great social pile; then the world of the termites and their fungus, communicating in the mound through waves of chemistry and water vapor; and then the world of the plants and geckos on the surface. Way up in the air, a giraffe obliviously munches on a tasty leaf. And from the air, a regularly ordered carpet of fertility and super-fertility becomes evident. And finally, a planet with an atmosphere.

Like the giraffes, we humans are ignorant of the vast churnings of smaller and larger worlds that we cannot see. We anthropomorphize or abstract these relationships into puny concepts we can understand: aristocratic insects, altruism, competition, cousins, bad guys and good guys. But these collaborative behaviors, and the sensing and signaling capabilities they require, may be the building blocks of complexity.

For a little while I had recriminating thoughts about the failure of humans to see beyond ourselves into the vast universe: we have so little ambition! But then I read a speech on the problem of scale in ecology, given by the Princeton ecologist Simon Levin. And when I read it, I realized that we are, as Patrick Keeling said, subjects in this experiment ourselves, and our fitful awareness is part of what makes us human. Levin said that the world needs to be studied on multiple scales of size, time, and organization—there is no one "correct" scale. And in fact, the scale at which we see the world is a product of how we've evolved, and how we will continue to evolve. "The observer imposes a perceptual bias, a filter through which the system is viewed. This has fundamental evolutionary significance, since every organism is an 'observer' of the environment, and life history adaptations such as dispersal and dormancy alter the perceptual scales of the species, and the observed variability." For humans as well as termites, these limits in how we perceive the world are the very core of who we are.

TWENTY

THE SOUL OF
THE CELL

CALIFORNIA

OVER THE EIGHT YEARS I followed termite scientists, many of them accomplished some version of what they set out to do. Sometimes the process was slow and halting, but other times— as with Kirstin's insights—things shifted quickly. Scott and Radhika and their teams continued to work in the field and in the lab, and they'd made surprising progress—for a field that was literally made of mud. Likewise, Phil's teams had eventually cracked the Rosetta stone question and then gone on to explore further elemental questions in termite gut genomes. Rob and Corina's team resolved a series of old questions. But by 2016, Héctor and JBEI still hadn't made grassoline at a price that could compete with gasoline. And over the many times

I visited him in his lab, I watched him and his team struggling with what Radhika had called "learning how to think," to replicate the termite's work of digesting cellulose.

A simple way to describe what happened at JBEI was that the model they were working on did not turn out to be economically feasible, even though their work had brought the price of lignocellulosic biofuel from an estimated $100,000 per gallon to about $30. But that didn't explain the more interesting parts, which were that they were battling with scale and, perhaps, with biology itself.

I first met Héctor's team in 2011 in a stark white conference room, like a set in a detective drama where the scientists successfully solve the crime. Héctor was wearing a dark blue hoodie from the Berkeley fencing club, and he bounced lightly on his toes as he prepared to sic his team of five on a metabolic map.

The map resembled that Tokyo subway plan that showed how the termite metabolized wood—with all the reactions hanging off the sides like trains to the suburbs. But this map was for the lab's pet *E. coli*, which had had cellulose-degrading genes inserted in it. The drawing contained a boggling level of complexity: tiny writing, chemical reactions, and little arrows. Just looking at it, my head felt like it was in a vise of stupidity from which it could not break free. Héctor felt bad for different reasons: "We can only see a hundred reactions here. And we need to see seven hundred or a thousand."

A truly useful metabolic map would make it possible for people without intimate knowledge of biochemistry to work on bioengineering. You wouldn't need to be highly educated savants like Phil and Natalia to be able to match up genes and metabolic processes, and you wouldn't need a room of postdocs manipulating different databases to visualize the whole thing in your head. "Once you can zoom in and zoom out of a map of the metabolism, memorization becomes irrelevant," Héctor said. So the map was a first step in changing the way the lab, and bioengineering in general, worked.

On a deeper level, the map was an expression of Héctor's conviction that synthetic biology's task was not so much the dumbing down of the microorganisms by simplifying them into LEGOs, but the "smarting up" of humans. By applying the abstractions that had been so successful in condensed matter physics, humans could understand what was going on without losing biology's essential complexity. Eventually, the tool would combine DNA from metagenomics databases with data on RNA, proteins, and fluxes, and weave it into a metabolic story encompassing the complexity of real life and complex ecologies. Done right, a map could even enable predictions.

When the team met, there was no existing database that could do this. A former Apple programmer led a tour: some databases had great images of molecules but no representation of the reactions that created them, offering just a snapshot of frozen time. One combined a sophisticated database with an unstable interface and irrational displays. "Cats and dogs live together and the walls bleed," the programmer said, combining references to *Ghostbusters* and *The Amityville Horror.*

The database was a big challenge, but it was only part of the problem.

Used to working alone and being judged by their successes, biologists in other groups initially saw little point in the extra work of uploading their lab results—successful and unsuccessful—rather than just recording winners in their black books.

In the years that followed, Héctor alternately empathized with their reluctance and fumed about it. He deduced that most of the scientists used more sophisticated computer interfaces when they played video games at home. But many of them simply didn't see the point in doing so at work because their education had emphasized the hands-on work at the bench—like counting worms for a whole year. "It's too abstract," Héctor said one day. "People want to know how it relates to their daily grind. Most of these people have restricted themselves to knowing their [metabolic] pathway. They work with their hands at the bench, and this is abstract." I thought of Shaomei and her headaches and wondered whether it would be a relief to leave the bench or not.

When I later talked with JBEI's leader, the synthetic biologist Jay Keasling, about this process, he saw it as transformative—and he wanted the scientists to get away from the bench altogether. "From a practical point of view a lot of money goes into biology, and it's still a lot of people standing at the bench manipulating DNA and transferring liquids around. If we could take these time-consuming things out of people's hands, they'll have more brainpower and more time to spend thinking about things. We ought to robotize the benchwork so that it doesn't take so long to design an experiment and get it under way. We could save a huge amount of money." Over the years the lab created a robotic experimenter that could do some repetitive lab tasks, getting reproducible results. There was a possibility that biology could transcend the repetitive work at the bench, but what they called the voodoo issue remained.

Héctor's group ended up accomplishing all of the tasks he set out to do—making predictions about reactions, getting the scientists to save their work, and making the metabolism intuitive. By 2015, they had metabolic maps that I could understand—with gorgeous thick arrows to indicate the relative quantity of carbon flux in each reaction, and red writing to the side with absolute numbers. With only a basic sense of biochemistry, I could explore the extraordinary sequence of reactions that a single, not particularly bright *E. coli* cell does when it's hanging out digesting cellulose with borrowed genes.

And with it, I realized that the thing I'd been taking to be the center of the reaction—the Tokyo subway line of it—was actually a citric acid cycle, while the formation of fatty acids that could become fuels— the really important stuff—was off in the suburbs on the right side.

Jay saw the map as a win in the long struggle against the quirkiness of biology and its aversion to rules. "Someday I'd like not to worry about every detail. I'd like to be able to operate at some higher level." Later he said, "We're not there yet with biology. It's a long way off. Well. I don't know about a long way. But it's OFF. That doesn't mean we shouldn't try." To the naysayers who say biology won't let us get to the higher level? "I would say try. Try! You might get partly there."

In its first seven years, JBEI and the two other government biofuel labs put out nearly nineteen hundred peer-reviewed papers. There were many important discoveries. Groups worked on new ways to break down wood with ionic salt solutions. Another worked on a molecular foundry that assembled long strings of DNA, going from ten genes to one thousand in a short time. The gene fabricator herself was a biologist in her early twenties named Sarah Richardson, who was so thrilled by her work that she came into the lab on Sundays. She used her love of computer games and logic puzzles to redesign gene segments with custom nucleotide bases (other than A, G, T, and C) so that they could not replicate outside a test tube. There was the robot doing experiments, and there were teams doing what previously seemed impossible.

I mention all of this because JBEI was, and saw itself as, an exciting place to be. New rules and new roles were being made here. It was the kind of place where people came to work on Sundays and were happy. When I mentioned the artsiness I'd seen in Phil's jazz sessions, Jay became enthusiastic. He and his husband, a theater director, often talked about the role of creativity in both of their jobs. "There's not a lot of difference in creating a painting that's beautiful and developing an experiment that solves an unsolved question—they both take a great deal of creativity to get to the bottom of them." And yet, his children didn't want to be scientists. "I guess maybe I've been a failure. I guess maybe they think I work too hard and it's too boring." Nobody thought that way at JBEI, though.

JBEI'S QUEST for cheap grassoline was complicated by circumstances outside the lab.

When the project first started in 2008, the price of oil was very high, U.S. dependence on oil imports was a concern, and there was some political consensus on reducing greenhouse gas emissions. And then the price of oil went nuts between 2008 and 2016, ranging from $147 per barrel to $28 and every price in between. Even if JBEI did hit its production targets, there was simply no way biofuels could compete at the bottom of the price zone. In addition, as the United States started producing more domestic oil from shale, the political climate that had been willing to underwrite a green technology initiative evaporated. Climate legislation now seemed impossible. Héctor and his team went to brainstorming sessions to learn how to be more entrepreneurial.

Héctor was always energetic, but some years he seemed down. More than once he said he was banging his head against the wall. The shape of his beard still featured a perfect symmetric point, but it never was as stark and angular as it had been the first time I met him. I haven't seen him bounce on his toes for years. I wondered whether he'd stay at the lab or go elsewhere, but he powered onward, gradually adding more people to his team.

When Héctor was feeling discouraged, he emailed himself encouraging quotes about the nature of science. Typical was one from Sydney Brenner, the scientist who discovered messenger RNA but had to wait years for his colleagues to acknowledge its existence: "What people don't realize is that at the beginning, it was just a handful of people who saw the light, if I can put it that way. So it was like belonging to an evangelical sect, because there were so few of us, and all the others sort of thought that there was something wrong with us."

The stress of keeping the faith was evident when I watched Héctor's team hold a meeting about flux analysis in late 2013. By measuring the movement of carbon—the flux—through the cell, with the help of sophisticated statistical analysis, it was possible to figure out where carbon was going in each reaction. Héctor's group hoped to find bottlenecks in the system and provide extra carbon at particular stages of the process to goose production.

The room was already dark when we entered, and the team's faces were only half illuminated by a large slide of a metabolism diagram—the subway map—projected on the wall. Recently the team had forecast that a tweak might yield a 40 percent increase in fuel production from the E. coli. A sleepy and slightly distracted postdoc announced that he had achieved a 100 percent increase.

Héctor was not happy. Instead of doing the standard reaction time of twenty hours, the postdoc, looking for increases, had left the microbes in solution for eighty-four hours to get that 100 percent increase. He had committed the crime of cooking: there was no data to explain what had happened over the final sixty-four hours. Héctor was exasperated, and he told the postdoc to go back to the lab and mind the method, not the results, noting, "Otherwise we'll go down the slippery slope of the last five years." It was too dark to read the postdoc's face—or anyone's, for that matter—to see how they felt about Héctor's frustration. Maybe they had meetings like this every week. Héctor ended by giving a pep talk, saying that there was no excuse for systems biology to have tools from the twentieth century.

After they left the conference room, the team went back to their cubicles, which were next to a glass wall that separated them from the hallway with the green windows. These dividers provided no privacy, and the only decorations were their bike helmets. I spent an hour talking to a young postdoc who was running Monte Carlo simulations—a computerized calculation tool that substitutes random numbers for real ones to understand what outcomes are likely, and how likely each possibility is. The team hoped to integrate this with flux analysis to get a more accurate understanding of what was happening in the cells than the linear models. The postdoc said that trying to second-guess the *E. coli* was "like trying to nail Jell-O to the wall."

In the midst of our conversation I looked down at my hands and wondered why they were bruised. They were a strange purplish color, like I had a sudden case of blood poisoning. I held them up in front of myself as though they were someone else's hands. And then I remembered the green windows: no longer an optimistic lens outward on the world, they'd become a tinted aquarium, holding the young scientists to a task that was getting murkier as it went along.

EVEN THOUGH JBEI'S DEEP GOAL was to enable biological engineering, finding molecules became key to the lab's survival. Exotic specialty chemicals, referred to as MultiFuel Biofuels, or MFB, with higher and more stable prices nudged cheap grassoline out of the way. (The scientists called them "More Fucking Biofuels.")

One such MFB was limonene—a lemon-scented solvent that is normally made by squeezing the skins left over from orange juice processing. It could be used as a fuel or an industrial ingredient. The lab figured out a way to make it with *E. coli*. Another one of the exotics was called pinene—originally made from pine and at one time an ingredient in the original Pine-Sol formula. Pinene can be combined with another molecule to create JP-10, an advanced rocket fuel that goes for $25 a gallon. Producing very high-priced chemicals for the military was one way to keep the lab alive long enough to find other biofuels.

This change mirrored a transformation in synthetic biology funding as a whole. Between 2010 and 2014, the Defense Department's research arm DARPA ramped up syn bio spending from zero dollars to $100 million, while investments by the National Science Foundation fell off. Only a few think tanks noticed, but the military had become the most influential funder in a field that started with a creed of "responsible innovation."

Meanwhile, the field's most noble molecule, the malaria drug artemisinin, supported by millions from Bill Gates, failed to change the world. The precursor drug that Jay Keasling's other lab had engineered took the place of wormwood, an expensive crop grown by farmers in Africa, India, Vietnam, and China. When synthetic artemisinin came on the market, the price of wormwood fell dramatically, making the real stuff much more economical. Given that the goal of the artemisinin project was to provide malaria medicine to the poorest of the poor, the fall in prices would have been a positive outcome if it had been passed along to buyers of the medicine. But drug manufacturers kept the medicine's price stable even though the cost of the main ingredient had fallen. In 2015 no synthetic artemisinin was produced, and Sanofi sold the manufacturing plant in 2016.

During this time, most of the public discussion of synthetic biology wavered between visions of utopia and those of apocalypse. The most successful products of synthetic biology were far more ridiculous— and telling—than that dichotomy allowed for. One was Genomatica's 1,4 butanediol (BDO), used in making Spandex and plastics. It apparently moonlights as a psychedelic drug. The field's legitimate block-buster was DuPont's 1,3 propanediol, used in creating polyester, paints, and glues. Produced by a genetically altered *E. coli* that lives on corn syrup, by 2021 it's expected to have sales of more than half a billion dollars a year. Both appear to be significantly better for the environment than the petrochemicals they replace.

Neither of these was going to save the world, but they did do the neat trick of turning corn syrup—often blamed for making us fat—into Spandex that might make us look a little thinner. There was something ironic and Darwinian (in the old "red in tooth and claw" sense) about Bill Gates's responsible molecules failing in the marketplace while chemicals to power missiles and make our butts look sexier succeed.

BUT WHY WAS progress so slow? When I first started reporting on JBEI, in 2008, scientists talked regularly about booting up yeast and bacteria with new DNA as if they were computers. It was a convenient metaphor for talking with the press. But it was a flawed one. Five years in, Jay Keasling and three other prominent synthetic biologists mused in a paper published in *Cell*: "An open question is whether biology is genuinely modular in an engineering sense or whether modularity is only a human construct that helps us understand biology." They questioned whether abstraction was "a useful tool or a necessary evil. . . . Biological systems, in contrast [to computer hardware], have been created by evolution and are not necessarily abstractable in ways that the human brain has evolved to handle." Perhaps the question wasn't whether we could reengineer *E. coli* to make grassoline. Maybe it was whether reengineering biology to the point where we could understand it would do us any good.

The complexity in the labs' test tubes suggested that the cells themselves had an agenda. As Héctor put it, "What we're doing is taking a bug [like *E. coli*] with no interest in producing biofuels and forcing it to produce them by inserting a pathway in there." The bug's "interest"—whatever it was—resisted manipulation. Eventually JBEI scientists learned to disrupt the cell's internal communications, or at least jam them, to keep the cell off-kilter.

The multiple ways that biology resisted engineering reminded Héctor of Carl Woese, his biologist/physicist inspiration, who had observed that, unlike an electron, a cell has a history. The engineering teams recognized that cell metabolism has memories that do not reside in DNA, but in some other network or way of storing information within the cell. "We have to wonder whether this memory is an emergent phenomenon, and maybe the DNA just produces proteins," Héctor surmised. Perhaps these tiny bags of chemistry "knew" or even "remembered" things about their environments. Their whimsical resistance to producing grassoline resembled—in a remote way—the quirky, idiosyncratic responses of the termites in the roboticists' petri dishes.

Whenever I visited Héctor, I felt the weight of his frustration. He was not waiting for Carnot so much as trying to conjure him: Physics Hulk smashing into the problems with great force. Just before his database of lab experiments launched in 2014, he reflected that it was "all this push push push. And then there's a critical mass of software bugs that get fixed. And then it's easy." And it was probably this conviction that kept him talking with me.

And in fact, this work paid off. By 2016, the team's work increased the output of fatty acids that could be used as fuels from that strain of *E. coli* by 40 percent using a systematic approach that could be applied to other problems. And the metabolic map tool combined with protein databases had increased production of pinene by 200 percent and limonene by 40 percent. They weren't anywhere near Craig Venter's dream of a million percent, but they were ramping up.

Yet the big question of how the termite's gut was different from a five-hundred-gallon steel tank was still out there, and it was standing in the way of getting the biofuel the scientists needed. Once the lab got one of their "bugs" producing a chemical, scaling up a thousandfold— from a flask the size of an orange juice glass to one the size of a kitchen garbage can—production would crater. How did the "bugs" change their behavior? And why? If there is a meaning in the scale and relationship of one organism to the whole—as Corina's work showed in the fields—it wasn't yet known in the bioreactor.

This led to greater uncertainty: fail to mix a bioreactor evenly and they'd end up with uneven streaks of oxygen and glucose that could create four-hundred-fold changes in production—making it a black box within a black box.

Seven years into the project, the idea of the cell as a machine, a computer, a factory, or a LEGO had given way to a more basic respect for its mysteries. In 2016 Héctor, Jay Keasling, and three others published a paper about what they'd learned so far. When I read it, I was pretty sure that this sentence had been written by Héctor: "Whether biological systems can be understood as a mechanistic composition of physical parts is a fundamental philosophical and scientific problem, epitomized by the understanding of the brain and the emergence of consciousness." It reminded me of Richard Dawkins's response to reading *The Soul of the White Ant*, when he wondered if his own brain was full of termite-like mindlets putting on a facade of unitariness, preventing him from knowing his own complexity.

Héctor had been promoted to a private office at JBEI when I asked him about that note. He laughed. "I had to fight a lot to get that sentence in there. But can we understand the brain from looking at the different parts together, or do we have to invoke some soul or something like that? Something metaphysical? That's one of the big questions of biology in this century. However, most biologists do not even consider this a question."

TWENTY-ONE

EMPATHY AND THE DRONE

NEW JERSEY

ONE NIGHT, PROCRASTINATING OVER SOMETHING, I obsessively Googled the Coronado National Forest, where I had gathered termites on my first safari with Phil in 2008. I've forgotten what I was looking for, but I eventually learned that while I was searching on the ground with my aspirator, as many as three Predator drones had been flying above me, controlled by pilots in air-conditioned trailers back at Fort Huachuca, about twenty-five miles away. Predators look like the offspring of an airplane and a mosquito: because they're unmanned, their bodies are lightweight, and their wings are long and spindly. They have no windows, so they appear eyeless, but from a height of twenty-two thousand feet they can "see" whether a person eight miles away is carrying a backpack.

So I had not been alone on my first termite safari. Via joysticks, drones, and video, the landscape had been engineered into something like the termite mound: a zone of total sensory awareness, a space tuned to detect invaders by way of complex feedback loops rather than conscious thought. Comprehensive but uninterested: after years of being consigned to places like Afghanistan, Yemen, and Pakistan, the drones had come home. By 2014, ten of them were flying along the border.

Without any democratic discussion, we'd made a choice to have more control over space via a filmy, unthinking, constant presence that is one of the essences of termiteness, the very soul of the technological superorganism. The definition of the human, Freud once said, was that we'd never give up our individual liberty for the group like termites do.* But by becoming a bit bug, we are losing some of the essential qualities that make us human.

REVEREND BILLY'S WORDS about vetting the RoboBee through the democratic process had stayed with me: "We can't let this be decided by the military." Over the years that I followed termites in the lab, what started as a somewhat goofy romp through the desert in search of earnest solutions had become a militarized space. Everything termites do, the military would like to do, too. It was not a coincidence that JBEI was making molecules with military applications while the U.S. Army had funded Scott's work on cognitive traps, and various versions of swarming and flying robots. DARPA was even funding a program to engineer live insects as "allies" to deliver protective genes to crops via viruses.

This investment isn't specific to insects. The military funds a tremendous amount of innovation in the United States. The Internet, GPS, Velcro, the turbines that turn natural gas into electricity—all grew out

---

*"It does not seem as though any influence could induce a man to change his nature into a termite's. No doubt he will always defend his claim to individual liberty against the will of the group. A good part of the struggles of mankind centre round the single task of finding an expedient accommodation—one, that is, that will bring happiness—between this claim of the individual and the cultural claims of the group; and one of the problems that touches the fate of humanity is whether such an accommodation can be reached by means of some particular form of civilization or whether this conflict is irreconcilable."

of military technology. The military's role as an inventor and investor makes it a complicated Santa Claus for both scientists and consumers like me, who can now protest war while wearing Velcro sneakers and tweeting our indignation on ridiculously cheap smartphone computers. The issue of why military funding for technology succeeds while Bill Gates can't buy a molecule for malaria sufferers is a different book. And, maybe ironically, military support for technology does have one advantage over a purely corporate or market-based approach to adopting technology, in that citizens have protested and even worked to get bans on weapons that we considered immoral in the past.

It's easy to see why the military would be attracted to the idea of becoming armed bugs: they can be both small (as bugs) and massive (as swarms). The Berkeley roboticist Stuart Russell writes: "As flying robots become smaller, their maneuverability increases and their ability to be targeted decreases. They have shorter range, yet they must be large enough to carry a lethal payload—perhaps a one-gram shaped charge to puncture the human cranium." The perfect assassin, in other words, is an exact replica of a truly functional RoboBee, but armed with that cranium-puncturing charge, a single-serving death bot.

And while these robobugs are not cheap now, there are great hopes that they will soon be cheap as dirt. In an influential white paper called "Robotics on the Battlefield Part II: The Coming Swarm," the RoboBee gets cited as evidence that soon the Defense Department could use 3-D printers to make a billion drones at a dollar apiece, producing "smart clouds." The author proposes that these tiny robotic killers could "flood" civilian and combat areas to find enemies, unconsciously echoing Von Trotha's "rivers of blood and money." The prospect of one-dollar swarms appears to be the perfect abstraction of power, removing all the limits of economy, power, and the political risks of losing the lives of soldiers.

The swarm is already with us. In late 2014 the U.S. Navy began testing swarms of robotic speedboats. By 2016, a cloud of 103 small drones demonstrated "collective decision-making" and "self-healing" in the skies above China Lake Naval Air Station.

As I've worked on this story, I've been surprised at how often—and how eerily—echoes of the time right before World War I come up. In 1904 Von Trotha turned his machine guns on the Herero in Namibia, and in 1909 Fritz Haber invented fertilizer and the military commercialized production quickly.* At the time, no one could grasp how these huge technological leaps—industrializing killing and farming—would affect the world. It was a moment when we went over a technological waterfall, with no going back. Billions would live from the new supply of food, and millions would be killed by the new weapons. Utopia and apocalypse arrived at the very same time.

I keep returning to *The Glass Bees* because of the way the author divided the realms of technology and morality. New technology may enable new behaviors, but only the changes in our moral universe permit us to do the things the technology allows. When the cavalry officer in *The Glass Bees* sees the moral centers of his world—chivalry and honor—fall away, he is changed. Just as the innovations of 1909 brought us poison gases and nitrogen fertilizer on top of the machine guns already in use, we're now at another time where our ability to both kill and dramatically change the world around us is leaping ahead, but we haven't given it much thought.

The popularity of the RoboBee as a military concept demonstrates that one of the oldest tactics of war—dehumanizing the enemy—is being abandoned. In Namibia, Von Trotha used "extermination" and "Cleansing Patrols" to suggest that the victims of the genocide were not people but vermin or insects. The swarming drone turns that old

*Fritz Haber converted from Judaism to Christianity to become a German citizen. As World War I began he wanted to help in the battlefield, but German generals were not initially interested in the poison gases he had formulated. Echoing the debate over machine guns, General Carl von Einem described it as "very unchivalrous," used only by "blackguards and criminals." Haber began experimenting with chlorine gas as a weapon, even going so far as to put on a uniform, design the delivery canisters, and travel to the battlefield in Ypres to figure out when the winds would be right. The first time it was used, it asphyxiated five thousand soldiers in minutes, wounding thousands more. Haber's wife, an accomplished chemist who had warned him that he was perverting the "ideals of science," took his service revolver and killed herself. After the war, poison gas weapons were banned and Haber was vilified as "the father of chemical warfare." He died before seeing another substance he'd helped to invent, Zyklon B, put to use in death camps. Mustard gas later became an important chemotherapy drug for treating some cancers.

paradigm on its head: we are now presenting ourselves to our enemies as technological insects, dehumanizing ourselves. The intent may be to show total technological domination, but we will also demonstrate that we do not value our shared humanity.

I asked the political scientist Peter W. Singer, who writes about technology and war, about what it means to present ourselves as insect drones in faraway cultures. The risk of autonomous weapons is that we appear not just inhuman, but also inhumane, which is fine if your sole goal is to create fear, he said. "But it backfires if you're trying to get the swing vote in a conflict. With terrorism, the only way we win is if we win the majority. An insect could be the wrong weapon for that task."

I WONDERED WHETHER anyone was thinking about the impact of those billion swarming drones. Two people told me to find Mark Hagerott, who was then deputy director of the Center for Cyber Security Studies at the Naval Academy.

I caught up with Mark at a cyber security conference at Rutgers. The RoboBee is on the first slide of the PowerPoint presentation he's been refining for a decade. An antsy navy captain with freckles and a graying beard, an open expression, and an easy laugh, Mark tossed and turned in his chair like a teen while we talked after his presentation. After tours in the Persian Gulf and Afghanistan, he had recently returned from an arms control conference in Europe where he had argued—to mostly deaf ears—that autonomous (self-controlling or swarming) weapons should be strictly limited by treaties. At the core of his argument is the question of human empathy.

The way he sees it, we're crossing two important thresholds at one time—machine-to-machine combat, and software (such as swarming algorithms but also including GPS and guided weapons systems)—that are beyond human comprehension. The mixture of the new machine realm and the acceleration of "Arthur C. Clarke" has put us at what Mark calls an "epic" moment where the paradigms of war are changing so dramatically that current laws and norms no longer hold—not unlike what happened in the time just before World War I.

To understand why this is so, he says, we need to think about the history of war. Early humans fought hand to hand, then moved on to throwing rocks, using bows and arrows, and eventually to catapults and cannons. Even as ships got more sophisticated, a battle consisted of ramming each other and boarding troops to fight hand to hand— people were still clashing with other people.

When the Spanish Armada set off to fight with England in 1588, they were surprised by redesigned British ships—lighter, easier to control, and armed with heavy cannons that defeated the Armada from a distance without boarding. There, Mark says, we crossed a threshold from human–human fighting into an integrated human–machine realm of fighting that grew more and more sophisticated with machine guns, submarines, and tanks, and reached its nadir in the flying aces and kamikaze pilots—men enmeshed with their machines like humans with prosthetic wings. It has continued through guided missile systems and the rest, because there is always somewhere a junction between a human finger and a machine's trigger. In 2001, when the first semiautonomous drone showed up, we began to slide into a third realm of war, in which machines combat other machines.

For Mark, what's crucial about this third realm is that between the machine and the complexity of the software, it's no longer possible to find that crucial place where the human finger meets the trigger. And this brief fleeting moment before the two meet is where empathy happens, or doesn't happen.

Empathy is embedded in the Geneva Conventions' Laws of War, where battlefield killing must meet three criteria, or rules of engagement: It must be necessary, the person pulling the trigger must distinguish between combatants and civilians, and it must be proportional. Each of these judgments requires a level of humanity from the warrior.

Empathy, in Mark's telling, is what keeps democratic governments democratic and puts despotic ones in check, no matter what weapons are at their disposal. For example: in 2011 the soldiers at Tahrir Square stopped shooting protesters because they knew that sooner or later they were all going to have to face each other. Without empathy—an amalgam of sympathy and self-interest—democracies may erode; the relationship between population size and military power will be upended; and what morality there is in war will shrivel up. Not necessarily in that order.

What he describes is not a robot apocalypse, but something much more human: a power grab. "I don't buy into the singularity or the idea that robots are going to take over, but I do believe that swarms of technology can allow despots incredible power," Mark told me.

In 2012 the Department of Defense decided that there had to be a "human in the loop" when drones and other autonomous weapons are used, but the relevance of that as a technicality is wearing thin. Drones permit the killer and the killed to be a great distance from each other, and all meaningful limits start to fall away. Mark offers two imaginary scenarios. In the first, a swarm of nanobots has a "lawful order" to execute combatants in a region based on whether they have certain DNA. A human may give the order, but there is no space for empathy to act. In the second case, a human is in control of 10 rifles or 100 rifles. "Does the argument change? What if your sniper is controlling 100,000 rifles and the targets are just green dots labeled 'valid target'? Is that meaningful control?"

Mark's talk of empathy reminded me of standing in the Harvard lab and feeling sorry for the Kilobots who limped along after their swarm. It seems humans are hardwired for empathy; and more than cognition, empathy may be what *makes* us human. And we've been generous with it: over the past century, following Eugène Marais's lead, we've grown a surprising new empathy for creatures we once feared, like gorillas, tigers, and polar bears.

Removing the messy humanity from our computers seems reasonable enough, but removing the humanity from our weapons systems is of a different order. Some theorists have proposed that autonomous robot armies will make better soldiers because they won't be racists, or get drunk, or be mean, or have bad days and bad judgment and bad aim. But to Mark, the very essence of war is the human element. His experience in Afghanistan showed him that policing, stabilizing, and nation-building require empathy. These fights, he says, are often about people being preyed upon by their governments, which can only be addressed by a bunch of frazzled humans, not a swarm of bots.

In the course of our conversation, Mark ended up taking both ends of the debate—for autonomous weapon development and against—angling from side to side in his chair as if he were a one-man Punch and Judy show. He wants to inspire regulation to keep autonomous weapons out of human spaces. Having looked at the potential uncontrollability of those billions of tiny swarming soldiers, he advocates limits: limits on numbers, requirements that they be no smaller than a human, limits on fuel sources, and limits on human impersonation.

On the other hand, he believes it would be irresponsible not to build them and hold them as deterrents. And he sees autonomous defensive weapons in nonhuman domains like deep underwater and in space as a necessity. As he articulated each side of his argument, he torqued more violently in his chair. "It's probably unstoppable," he said at one point. "Let's face it. You're going to need a drone to police a drone. So I'd invest in defensive drone technologies. Right?" A series of impossible moral choices.

Mark's ultimate message is that the sooner we have this discussion and the more we can reach some rules about the use of autonomous systems, the better chance we have of achieving more positive than negative results. Though the European Union, the UN, and other organizations (including the U.S. military) have been discussing governance of lethal autonomous weapons systems (called LAWS) since 2013, these discussions are still in very early stages because it's difficult for most countries to imagine the full implications of autonomous warfare, or even what the weapons could be.

We do have a history of limiting some weapons. In 1899 more than two dozen countries met in The Hague to discuss regulating the use of balloons and projectiles for spreading poisonous gases, as well as hollow-point dumdum bullets. The United States did not sign. Teddy Roosevelt pressed for the second Hague Convention in 1907, and this time the United States and the United Kingdom signed a treaty prohibiting dropping bombs by balloon. But it wasn't until World War I that it was clear that the industrialization of the tools for killing—machine guns, gases, bombs—had changed everything. After the war, the Geneva Conventions put further limits on the means and methods of killing because of international consensus that the technology for killing had outpaced our moral imaginations.

Since then, treaties have banned blinding lasers, neutron bombs, and chemical and biological weapons—often with the assistance of scientists in those fields. Of course there are cheaters and stockpiles, but the fact that we're all sitting around discussing LAWS policy instead of having been blown to tiny bits during the Cuban Missile Crisis or some other moment of tension is a testament that the combination of human empathy, treaties, and enforcement does a decent job.

WHEN I WAS visiting Radhika's lab at the Wyss Institute, I talked with a postdoc who told me he worked on the RoboBee using a programming language called KARMA. I thought that was a funny name for the language, given the military applications for this kind of tiny robot, and I asked him what he thought about the possibility that his robobugs would be used for combat or surveillance. "I honestly feel a little ambivalent," he said. "I feel my job is innovative technology, and whoever it goes to makes a decision of how it's to be used. Whether it's good or bad, we can't stop the technology because it might be used for bad." This is a fairly typical sentiment—and an understandable one—from builders of robots, and also from the scientists working in synthetic biology. They see themselves as engaged in a variety of hard problems that require solving, and grappling with the eventual application of these solutions is simply too much given the other things they're juggling—family, funding, and the stress of academic production.

I understand why scientists wish to keep their heads down and avoid politics—which are certainly vicious and irrational and frequently have a zombie-like antiscience bent—to get their projects out of the "toy" stage. While individuals like Radhika are outspoken about the current limits of swarming technology, the field is very much focused on the future, which often means emphasizing potential successes while downplaying possible complications. It's also true that scientists have no control over the uses their inventions are put to when they leave the lab. One can invent an insecticide and expect it to make people's lives better, without being aware that it will be used to kill millions of people.

Discussing the possible applications of developing technology is sometimes actively discouraged as "politicizing" scientific disciplines. A military policy expert who spoke about ethics and drones at an engineering school told me he got a critical email from a professor afterward for even bringing up the subject. At a national meeting of synthetic biologists in 2015, I watched as several scientists dismissed the value of an MIT program to anticipate and avoid real-world problems with their inventions, which included such worrisome products as synthetic opioids. Generally, the scientists echoed the conclusions of President Obama's panel on the synthetic microbe Synthia—worrying that any controls or regulation would hold the field back.

Even though the risks of both synthetic biology and autonomous robots are unknown, very little high-level attention is given to them. Less than 1 percent of the federal money invested in synthetic biology goes toward understanding risks. And the study of the social impact of autonomous drones is even smaller. This is an oversight, because when we talk about technology as purely innovation, we fail to anticipate how its use evolves once it enters the world in large quantities—as with machine guns and fertilizer.

Out of necessity, scientists may see themselves as termites involved in creating their pile of mud balls. In their individual experience they cannot see or understand how the whole mound they are building will or could work. By sticking with the local, they leave the global out of sight.

In trying to understand how this plays out in society, I read an article by the ethicists Sheila Jasanoff, J. Benjamin Hurlbut, and Krish-anu Saha. "Trust in science is just as fragile and just as much in need of regeneration when science, in effect, takes on the tasks of governance by shaping society's visions of the future," they wrote. "The challenge for democracy and governance is to confront the unscripted future presented by technological advances and to guide it in ways that synchronize with democratically articulated visions of the good."

Following termites led me to the much larger question of who gets to imagine and define the future. And there is no simple answer. Some scientists in some labs have the ability to pursue technologies that they think are possible for problems that they think should be solved. They make some decisions about how these solutions will work: what scale they operate at, who controls them, how they are funded. But, as the failed introduction of artemisinin showed, even they lack the power of big market movers and the military to determine which innovations succeed. To borrow a concept from the termites, there are influential individuals, but no one seems to control the process.

How could we decide, democratically, what we'd like the future to be like? The coming decades will require many choices about how we interrupt the processes of the Earth, the atmosphere, and one another. We have to create institutions that can reckon with the moral implications of these choices now rather than simply waiting to see how things turn out.

One thing that has become clear to me is that those of us outside of science have a responsibility to better understand science and to consider the implications of technologies. By talking alternately about apocalypses and saving the world, we participate in a game of mystifying technology, turning it into a sort of play, an imaginary space, or an entertainment that we enjoy passively. We are preventing serious discussion before it even happens. We need to call technology what it is—an abstraction of power, politics, and economics. And then—if we are going to take ideas from the termites into our human realm—we should use them to become more human, not less.

TWENTY-TWO

WHITE ANTS

AUSTRALIA

I KNEW OF ONE PLACE where the citizens had decided on a "democratically articulated vision of the good," and set a scientific agenda for more than three decades: the Yolngu community where Dieter had worked. In 2014 I returned to Australia for a social insect conference, visited Phil's lab in Brisbane, and then flew on to Gove Airport in the far northwest of Australia, hoping to see the rehabilitated land at the bauxite mine. I rented an old car and drove a few miles to Yirrkala, a collection of about a hundred brightly colored houses perched above the blasting tropical blue of the Australian side of the Arafura Sea. I was staying there in a guesthouse at the Buku-Larrnggay Mulka Art Centre.

The next day I sat on the ground in the shade of the art center's tall trees and tried to explain to Yalmay Yunupingu, a self-possessed town matron, why I was writing a book about termites. Yalmay was kind and she nodded with warm amusement. "Termites," she mused . . . "White ants." She pronounced the *t*'s as though tucking in their corners. "Well, termites are a good metaphor, right?"

Yalmay, like every Yolngu person I met in the area, was a true cosmopolitan. She was a noted artist and teacher herself; her late husband had been an educator, an actual rock star in the band Yothu Yindi, and was named Australian of the Year in 1992. Yirrkala is remote, but it is by no means removed from the world: its artworks go to museums and collectors everywhere. In less than a week a few thousand people would fly in for the Garma Festival of Traditional Cultures, twenty miles to the north, where Tony Abbott had promised to be a "prime minister for indigenous affairs."

Yalmay introduced me to the idea of "white-anting." I'd heard someone else say the community had "spiritual anxiety" about the future. When I asked Yalmay about that, she said that Rio Tinto, the current mine owner, was just "white-anting" the place: eating out the center invisibly, and then abandoning it.

The accusation of "white-anting" turned out to be one of the worst things you could say about anyone here.

Getting to Arnhem Land wasn't easy. It required permissions and expensive flights, all complicated by the fact that I had very bad bronchitis. Dieter hadn't been able to come along. He had visited his work in Gove the year before and seen that his methods had been mostly abandoned. I considered not going, but it didn't seem right to merely report on the rehabilitation project without actually finding out how people felt about it now.

Put another way: I would have felt terrible if I hadn't figured out how to visit Gove, but now that I was here I felt rotten. August 2014 was a particularly bad time for the community and the mine. Six months before, the current owner had begun closing down the local aluminum refinery and laying off the fifteen hundred people—mostly from other parts of Australia—who worked there. Now twelve hundred mine employees had left the area. The town of Nhulunbuy, which had been created for the miners, was already folding up. Houses were empty, stores were going out of business, and the local dance teacher was preparing to leave. Having been forced to deal with the presence of the mine for forty years, the community was on edge about what its departure would mean. The only thing that was known was that the community would not get back the land the mine was on for another thirty-nine years.

Now that I was finally in Yirkkala, I realized that only a few people remembered the rehabilitation project. People who knew the land before the mine came were mostly gone, and the people who would appreciate the place after the mine closed were children now. The community had been effectively blocked from going on the rehabilitated land for eighty years, and so it seemed to have disappeared from their collective consciousness. The success that I'd read of in journal articles was something different in real life.

I HAD ARRIVED just before a Friday holiday, so I had three days when I couldn't do any "official" interviews. I hung around the art center and talked to people, and early in the afternoon someone said that a woman named Valerie had called asking for a ride to find a tree. Would I take her? Sure.

I found Valerie on the porch of a raised wooden house hidden behind a tangle of trees, painting with stone-ground red pigment on a piece of a tree. An artist who also makes and paints didgeridoos, Valerie Milminyina Dhamarrandji was shorter than me, with a silver pageboy, sinewy arms, and an air of energetic hilarity. She wore a giraffe-patterned top and a blue skirt with circles on it. When I introduced myself, she laughed and grabbed a cream-colored purse and an orange-handled ax, then called her daughter and two granddaughters to join us.

First, Valerie said, we needed to drive ten miles through the stringybark forest to Nhulunbuy to get supplies. The road was dirt. The forest had a typical rustling canopy of thin eucalyptus leaves and a park-like understory. The ground was littered with leaves, small shrubs, flowers, and grasses. The light felt golden partly because the intensity of the sun was softened by the trees and partly because of the reflection of the yellow straw from the floor. The seasonal burning was about to start. As we sped past tens of thousands of trees, Valerie spied individual ones deep in the light and commented on them fondly.

Valerie and I were only a few years apart in age. In her childhood Yirrkala was a small missionary settlement, and people lived a subsistence lifestyle, eating seafood, foraging, and hunting. I'd seen movies made in the 1960s and '70s by Ian Dunlop, a British-born filmmaker invited by the elders as part of a deliberate strategy to make their lives and culture matter to the rest of Australia. One of the movies follows a family on a trip to the beach, where a dozen children frolic in the ocean, fishing and digging for clams. Valerie might have been a little older than those kids at that time.

Valerie told me to stop for a quick walk in the woods. She started by pointing out the difference between the paperbark eucalyptus tree and the stringybark. The paperbark tree appears to have burst its seams, with layers of curling thin white pages appearing from gashes in the outer bark. The stringybark is covered with a fringe of woolly bark strings. Paperbark can be boiled and used for colds, she said. I prepared myself for a mini lecture on ethnobotany but we were quickly into some kind of cosmology, with a cascade of identifications, each leading to some new point in time and space. There was the yellow acacia flower, and when it's out the oysters in the bay are fat. The pandamus grass can be used to make a basket. And here, under the leaf litter, is a grass with bright red roots that can be used for dyeing pandamus for baskets. When a shrub with red waxy flowers blooms, the sharks are fat and ready to eat in a nearby bay. And when the stringybark eucalyptus flowers, the honey will be ready inside the trees. Valerie roared with laughter, explaining that every plant and animal and bit of land is affiliated with one of two moieties, or paired opposites, which are also tied to clans and to individuals. Everything here is relational to everything else and then interconnected, until the forest is a giant Internet leading to stories, lore, law, medicine, and fat delicious sharks.

There were other associations: the honey is related in some ways to the sea in the songlines and to the character Wuyal the "honeybag man," but she thinks I might be interested in it because the termites hollow out the trees where the honey is found. In fact, there were signs of termites everywhere—the little pointy hats and pinkish *Coptotermes* mounds crawling up the trees.

We got back in the car and continued toward town. On the way we drove over a bridge between two long red embankments. Underneath us drove enormous red-dusted mine trucks, as though there was a parallel land here, twenty feet below the one we were on.

The town of Nhulunbuy was built by the mine in the 1970s, and its grass-riddled grid reflects that era's stubborn belief that rational planning could avoid hassles. Australia saw itself as a mining concern doing business as a country. After the mine was built, the government started giving the Yolngu welfare payments. Forty years later, there are calls to cut that welfare on so-called moral grounds. Later Prime Minister Abbott said the country couldn't afford to subsidize the "lifestyle choice" of living in remote towns. Another promise white-anted.

We stopped at a BP station. Valerie smoked a cigarette and appraised me. Why had I come here? I started to talk about the rehabilitation, but she cut me off. "Oh yes," she said. "There was a man named Dieter. . . ." She had worked as a planter and she remembered him well. She happily recalled the crews she belonged to and the work they did. We discussed posthole diggers, which we had both used. There was a spot under the conveyor belt that she wanted to take me to.

On the way back she saw a good tree. With her ax in hand, she ran to it. It had a termite mound on one side of it, which could have been a good sign, but it wasn't, because the termites had bitten right out the other side. So she began to run through the forest, hitting all the trees larger than eight inches in diameter with the ax. She said that if they weren't hollow, they'd sound "flat." So *this* was the forest where eight out of ten trees are hollow.

She found one with good resonance and began whacking it with her ax. When she'd bitten through half the tree, and we could see that it was hollow, she put her arm on the tree like it was a friend and calmly looked up at it, waiting for it to fall. It did not. She did some more work with the ax. Ox-like, I offered to give it a big shove. It thunked over. Valerie chortled, saying it needed to be cut nicely with a saw, and would I do that while she went to get another tree?

I began to saw the tree into an eight-foot length, which wasn't hard except that under the sun and with the large quantity of prednisone I was taking for my bronchitis, I began to sweat as if I were a human-shaped rain cloud. The granddaughters had stayed behind to watch me. They asked why I was sweating and I said I came from a cold country. We talked as I continued sawing until I had a decent flat cut. When I stood up, I thought I might faint. Prednisone allows you to pull off the trick of being very sick and slightly superhuman at the same time. The girls thought the cut would be acceptable to their grandmother. I put the log on my shoulder and followed them deeper into the woods.

Valerie had cut down another tree in the meantime. She wondered about me sawing it. I said I couldn't saw anything until I'd had some water. Big pulsing neon spots were dodging through the forest.

I lifted and dropped her new tree segment until the termite comb fell out of it. We discussed the fact that it was some termite's house, laughing. I carried the logs back to the car, but they didn't fit, so we left them by the road and returned to Yirrkala. On the way back Valerie allowed that termites are bad but they also have good points. They hollow out the trees for honey, and make the logs she uses for sculptures and didgeridoos that she sells in Nhulunbuy.

One of the art objects that's made from termite-hollowed tree trunks is the larrakitj, or ceremonial pole. In the old days, larrakitj held human bones in their hollow centers, but now they are created and sold as art objects. I went to examine some at the art center. The poles are slightly shorter than me, and they feel like beings themselves. Many are covered in cross-hatching or other patterns that hide and reveal things simultaneously, like the sea lapping over a rock outcrop. One series was created by Wukun Wanambi, who formerly worked at the mine. He smoothed the bark of the tree trunks, painted them white, and then etched and painted thousands of anchovies on them. Where there were knots in the trees, he painted a bird's head so that it appeared that the bird was both eating the fish and being eaten by them. Staring at one of these poles felt like a long conversation.

Will Stubbs, the former lawyer who manages the art center, told me that the only way to understand this place was to be confused. "What's most important about the larrakitj is what's not there. The center eaten by the white ants, the unsaid things in the paintings, the things hidden by layers. It's an aesthetic, a law, religion, a taxonomic system, a science, and a mathematics."

I REALLY WANTED to see the rehabilitated land, but it was all controlled by Rio Tinto. I had written them letters asking permission, but when I finally arrived, the company's representative wrote that I could not see the land "at this time." That qualification made me laugh. When would there ever be a right time?

I drove over near the mine where the land was stripped of vegetation and the dirt was raw. Across the bay sat the aluminum refinery, a collection of rusting machinery and tanks with a cartoonishly retro air that seemed much more than forty years old. A wisp of smoke came out of it, but there were no cars in the parking lot. People had told me they feared that the mining company would keep the plant barely open like this to avoid having to clean up the site.

I visited the office of the contractor that did land rehabilitation for the mine now. A self-effacing white forester said they no longer used Dieter's system and the planting schedule was determined by a computer program. Tracking termites had been abandoned, but they did try to get the acacia to grow within five years and they still hired a few locals seasonally to gather seeds for replanting. We drove near one of the rehabilitation sites and looked at it from a distance. It had fluffy eucalyptus on top, a midstory of brush, and grass on the ground. To an untrained eye it looked more or less okay. The forester regretted that they were unable to burn the reforested land and as a result the leaf litter and the invasive species known as crazy ants had gotten out of hand. And so what looked like native forest from a distance was really a new hybrid.

IT'S POSSIBLE TO WITNESS the exact moment when the Yolngu elders communicated their wish that the land of the mine be returned to forest, because it was captured in a film made by the documentarian Ian Dunlop in 1974, called *This Is My Thinking*. The film starts in 1971, when Daymbalipu Mununggurr takes his kids to a beach and finds foreign miners there without permission. By 1974, his worries about the impact of the mine have been realized: Nhulunbuy's new bar (called The Walkabout) was contributing to a culture of drinking. Later he sits through a meeting with employees of the mining company where his concerns about the bar and the safety of the red mud pits that hold caustic tailings are dismissed.

Then a mine engineer brings him to look at the nursery of African mahogany trees, Egyptian cypress, and other exotic plants ready to be planted on the mined land. In the background a young, blond, square-headed man in shorts and tall white knee socks paces anxiously. "It'll look very much the same, just twelve feet lower than before," the official says chirpily of the rehabilitation. Daymbalipu said that this sort of replanting was unacceptable. "I've got nothing against your way of thinking, but the aboriginal people have problems with this. The eucalyptus belongs to this country." The community wanted the land rehabilitated to the native species, not exotic ones. I played the video again and realized that the antsy young man in knee socks was Dieter.

How, I wondered, had the elders reached an agreement about what they wanted the land to look like? Another Dunlop film showed elders meeting and talking for hours about what the land meant and what they feared. Several people told me to try to talk with Wanyubi Marika.

One afternoon I got a call saying that Wanyubi was ready for an interview. A big man wearing a blue polo shirt and shorts, he sat at a table on a porch at the offices of the Bunuwal Group, on a bluff over-looking the beach on one side and the art center on the other. The waves were crashing and there was a soft breeze. Wanyubi was impa-tient, jangling his leg. In photos of him when he was younger he smiles brightly, with dimples and curly hair, but now he frequently scowls. The son of Milirrpum Marika, the signer of the bark petition who later was a member of the land rights suit, Wanyubi inherited the responsi-bility of being one of the leaders of the Rirratjingu clan in Yirrkala. He is also a well-known artist and he runs the Bunuwal Group, which rep-resents the indigenous owners of the area, overseeing several busi-nesses, including one that contracts with Rio Tinto to truck in fuel and other supplies, providing employment and income.

In 1984 and '85, he worked on the rehabilitation project, collecting seeds and driving equipment to reapply the topsoil after the bauxite had been removed. He remembered it as a neutral experience. "The land is a bit changed. It's not normal or natural but at least they put the soil back." He said they'd achieved the goal of getting the forest back, but it's a no-go zone for hunting and burning, with invasive grasses and animals, and so it was not part of anyone's life. When the community gets access to the land, he'll be an old man.

I asked him how the elders had imagined the future and come up with a vision for moving through it. He began to educate me in a way that at first seemed meandering. The songlines, he said, start from the horizon of the ocean, with the clouds breaking and the sun rising and setting. They talk about individual trees and plants and animals both at sea and on land. They talk about the stringybark trees. "We see what's been sung in the sea and on land and that becomes how we man-age the land," he said. "But these feral [invasive] weeds are not in the songlines. The crazy ants are not nor the buffalo pigs or the coastal gnats."

I had seen Wanyubi talking about art and tradition in a video from an art gallery. Then, he had explained that it is the job of adults to teach the songlines to children so that they understand what should be where, and how the world fits together. And as time changes, the new generation has to be able to adapt the old visions and responsibilities to new ones.

The community, he said, is no longer struggling about land rights; it's now struggling over sustainable economic development. "We're in another world now." In twenty years, he said, people may have lost their own languages and speak English, or the colors may have shifted, but the essential vision needs to stay strong.

On the porch, he explained how to use knowledge from the past to think about an unknown future. "We have to balance the stream with one foot on either side." He held his hands wide—on either side of a stream. "It equals good governance. Leading in a good manner."

"For example, if you want to go to that island over there . . ." He pointed out toward a small island. "The water is wavy and windy. You need a good sail and a good skipper. And water to drink. Plan it properly before the journey. Calm water becomes rough water and you meet with a big cyclone coming in. There will be some obstacles. The elders need to go through that vision—long, hard, tiring—so that our young ones can have calm. That is good governance, a system of policy for people to stick with and abide by."

Wanyubi had a last comment that revealed how detailed those group discussions about the obstacles of the future must be. "When the mine closes, the climate change will have changed the rainfall patterns and the walls of the red mud ponds will fall down and the caustic will be carried into the bay, where it will damage the marine environment. We'll see what Rio Tinto leaves behind and we'll suffer with this country." We stood up.

TALKING WITH WANYUBI changed my mind about what I was doing here. Termites had brought me here, but what I was really seeing was how a community discussed what it wanted for the future, reached consensus, and then built the political and social capital to make sure it happened. The bark petition, the protests, the lawsuit had all built political power. The documentary crews, anthropologists, artworks, rock bands, and Garma festival gave them social influence well beyond their remote location. They made sure the mining company regrew the stringybark trees by adopting Dieter so he shared their goals. The termites, with their hexagons of fertility, were the very last piece in the project. What had started the rehabilitation was a shared moral imagination, an attention to the nitty-gritty details of what was important for this community.

The times in history when small, remote communities living in resource-rich areas get to determine which problems need solving are rare. It's so unusual that the *Journal of Responsible Innovation* contains articles with titles like "Constructing a 'Futurology from Below.'" I don't think the elders constructed from below; they seem to have had a good sense of the landscape from a great height and over a long period of time. But they certainly did construct a futurology as they imagined the best possible outcome for a bad situation—regrowing the forests even though there was no known means of doing that at the time.

And this brings me to the second thing I realized in Arnhem Land: success depends partly on who tells the story and when. People who study anticipatory governance are trying to find reasonable accommodations between social goals and technological possibilities, good and bad. But there is also an unreasonable truth that what we want in 1971 may not be what we want in 2015 or 2045. While nitrogen fertilizer looked fantastic in 1909, it looks different in 2019. It's not that the decision itself was wrong, it's that new realities are constantly forming—the Escher hand redraws the hand itself.

In today's Arnhem Land, regrowing the stringybark trees is a significant achievement in a major disaster; forty years from now it may seem prescient. But right now no one particularly cares. It's only in the careful transects and grids of scientific journals that the "success" of a project can be separated from the living stream of history of the people who use it.

ONE MORNING I went to find Valerie. She seemed surprised that I'd returned. We dropped off her relatives and then drove out to the mine. On one side of the road there was the ocean, pale and wavy blue, and on the other a delicate mix of shrubs and grasses, guarded by fencing and a RIO TINTO sign. Valerie said this was one of the red mud ponds, where caustic mine tailings had been dumped into lined pits. Then grass was planted on top of them, and eventually bushes took root. These were the ponds that Wanyubi feared would cause harm in the future. In the distance was a broad red embankment, like a stripe made with a coral-colored crayon across the bottom of the sky. A tiny yellow machine labored on top of this distant Mars, and we could hear a mournful industrial beeping.

We drove further to where the big conveyor crossed the road. In the distance we could see the fluffy canopy of the tops of the trees. Valerie said that locals had been proud of the rehabilitation in the early days. Nobody who lived nearby wanted to look at the dirt exposed by the miners. Now that she looked at the forest, she wondered why all the work had been forgotten.

We went onward to Ski Beach, where Valerie had a house. Until 2000, Valerie had lived in another place, called Woody Beach, where she could easily gather trees and fish and clams, but that settlement had been torn down, and she resettled into a shipping container with a porch here. Valerie said that she could not go looking for clams or trees or fish without a car. It was expensive and ridiculous to call a taxi just to look for wood. The meaning of the old subsistence ways—all the old foods and the old knowledge—had changed.

The towns of Yirrkala and Ski Beach have long been divided over money from the mine. In Yirrkala, with its beautiful art center, art and ecotourism seem like industries of the future. So the impulse there is to preserve the trees on the land when the mine is returned. But the residents of Ski Beach, who currently get 72 percent of the royalties from the mine, have plans to log the rehabbed land, start their own small bauxite mines, and raise cattle. The intimate damage the mine has done to the community is much harder to map than its environmental impact.

On the way back to town we passed a banyan tree sifted with pink dust. Valerie said that her grandfather was born under it. Munggurrawuy Yunupingu was one of the twelve elders who made the original bark petition, and when the mine engineers came with a bulldozer to topple the banyan tree, he grabbed an ax and stood in front of the tree to protect it. "It was scary then but it's funny now," she said.

Before her grandfather saved this tree, the miners had knocked over an even more famous banyan tree that was the embodiment of the "honeybag man" Wuyal. The community was devastated and sent another bark petition to Parliament in protest.

As we drove back Valerie lost her good humor. "They really raped the land."

They certainly had. This mine had cost the people who lived here a lot. I had nothing to say. We drove back through the alien pink planet she'd been left with and picked up her family. Along the way, under the influence of the prednisone, the low sun, and emotion, I couldn't see anything in the sun's glare and drove off the dirt road into the ditch. No one was hurt, but the friendly mood turned sour, and I felt that I had let Valerie and her family down.

What was I doing here? Had I come all the way to this remote town to cheer on a random rehab project when the rest of the place was irrevocably hollowed out? Was I sitting on my aluminum campstool and shedding a tear over my aluminum computer for the way a bauxite mine has contributed to the destruction of an environment and the lives of the people who live here?

Was I a white ant? Had I hurt people while claiming to tell their stories? I tried to count how many aluminum jets I had ridden to reach the conclusion that bauxite mines are bad. What was my responsibility, as a writer and as a fellow human? I kept returning to this long after I left Yirrkala.

A certain kind of technocratic optimist might look at this situation and declare that the problem is aluminum. To that person, the way mining destroys the land in remote communities around the world constitutes a clear argument for synthetic biology. Engineer microbes to grow some new hard substance—maybe a kevlar-like form of the chitin that bugs use for exoskeletons—and use that to wrap our computers and our airplanes.

That would be an easy story to write. But even the story of such a technology—appearing to deliver us from our old politics—could harm people here by making their complaints seem irrelevant.

WHERE BEFORE I might have seen such stories about solving problems as being *about* a specific technology, I now see just the story. Science, along with our hopes and dreams and fears of it, is one of the big stories—almost an art form—of my culture. Like larrakitj poles, these stories hide and reveal our ancient myths in new contexts.

If I had not ridden all of those aluminum jets to get to Gove, I would not have understood so clearly that regrowing the forest was not the same as justice. Justice is a much bigger goal than problem solving, and it's far messier. It's one thing to create a medicine for malaria—or, for that matter, unlimited hamburger or grassoline—but how do we do right by our fellow humans?

When I first went to the Arizona desert with Phil's team, I anticipated replacing the oil wells, the Hoover dams, the mines—all the grand mechanisms of control of the previous industrial revolution—with steel tanks that had a smaller impact on the planet. But that vision, with its abstractions of power, looks different to me now: Who builds it? Who decides? Just as societies are the organizing principle for termites, so they are for us humans. We shouldn't imagine that technology will deliver us from our mutual obligations.

Staring into termite mounds had turned me into a version of the Victorians, looking for meaning in there. Only instead of queens and utopias, I saw all the things we don't know: a representation of the world's boundless complexity.

PART VI

TWENTY-THREE

THEM AND US

NAMIBIA

I N THE SUMMER OF 2014, I went back to Namibia. As I drove
north on that too-straight highway with its ominous yellow warning signs, the termite mounds greeted me with comic stoicism.
Chasing a silly story about robots building on Mars had upended the
way I saw my own planet—and myself.

Via my obsession, mounds became everything that mattered to
me: The meaning of life. The key to the future. A parable about the
interplay between the organized narratives of stories and the multilay-
ered data and process that is science.

What were the mounds but the dirt's impression of the solar system, sculpted by a few million eyeless bugs? Squint just right and these

dirt fingers were a map of the universe. And yet, they were also the building blocks of our world: dirt ball by grass clump, termites make the terrestrial maypoles around which a good part of the planet's fertility spins.

More personally, mounds are the human brain turned inside out. Fittingly, we can only approach their weird glory by pinging stories— queens, paradoxical protists, all of our hopes and anxieties about our own society—off their smooth sides.

Leaning over at that 19-degree angle, they're an ongoing meditation on the complexity of dumbness and the dumbness of complexity. And while we're on that subject: I still have no idea what to make of the fungus.

I was lucky to find the mounds at the moment when we can still see them as old-fashioned superorganisms—that not entirely scientifically useful concept. Someday maybe they will give up the lessons in complexity that will shape our future from global to local; and perhaps the chemistry that turns us into *Homo lignivore*. Let me put this another way: Right now a termite mound is a thing, a construct of fungus and termites and natural history. Someday we will live in it, with all its symbiotic by-products, its paradoxes of abundance and control, and its peculiar self-organized construction. When we do, we will no longer be able to see it the way we do now.

In its stubborn tolerance for failure and ability to adapt, the mound is also a hopeful beacon in a biosphere that's changing rapidly.

But a soul? Nah. I still don't know that that's the right word for the mounds' mysteries. The bigger question is: How, with our limited knowledge and clunky, termite-mound-like brains, do we determine what the important connections are between the world as it is and the way we represent it?

I DROVE NORTH to Otjiwarango, and then to the new farm where Scott had moved his lab. Scott, Berry the bird expert, and Rupert the engineer were all there, along with Paul the entomologist and two physicists who were measuring gas and temperature flows in the mounds.

Every night was an interdisciplinary salon with the familiar oryx steaks and squash. There were vegetarian lentils for the physicists. Rupert talked about engineering and Berry talked about the desert. The physicists talked about something peculiar that waterdrops do.

Radhika wasn't there, but I thought of her every time the cows mooed. These cows had a low rumbling sound, and whenever we heard them lowing late in the evening, everyone would pat their pockets to see if they had a call on their cell phones. It wasn't quite a propagation of local effects, but almost. She would have thought it was funny.

Paul was constructing "ant farms" between two pieces of glass, which had allowed him to figure out the termites' rules for tunneling. If one termite was in the tunnel, it went straight. If so many termites were in the tunnel that they piled up, some would start digging a branch off to the side. So the pressure of termites in the tunnel influenced how much it branched. These termite highways looked very much like a drawing of human blood vessels, with big trunk lines branching to ever smaller capillaries. These tunneling tendencies multiplied by the millions underground, creating the patterns of fertility that Corina had seen aboveground.

And still, when I actually saw this underworld revealed, I was amazed. One day Paul melted a lot of aluminum wire with a blowtorch and dumped it into a termite hole in the ground. We spent hours carefully removing the dirt from the elaborate lattice formed by the aluminum tunnels. The molten metal had traveled through the soil in a staggered grid. Once it was exposed, we could see that what feels like barren scrub is really New York's Union Square Park, with multiple termite subway lines converging below us. Their architecture was distinct: when they hit something hard, the termites would build a spiral down (or maybe up) to continue in the same direction on another level. It looked like a 3-D Etch-a-Sketch, except for the spiral staircases: pretty much what I imagine life on Mars might be like.

In fact, termites may be the closest I ever get to a space alien, because they remind me how weird it is to be human and locked into our own scale.

To really "see" termites we have to step out of the rules of being human. Speed up time and you can see that the ground beneath us is moving as many billions of tiny termite feet carry grains of dirt. If I could just situate myself in the correct time signature, I'd feel I was on something akin to a very slow-moving walkway at the airport. And if I could shed our timescale altogether, I could watch the termite mounds slouching across the landscape, following water, like those hybrids of dirt and bug that Eugène Marais predicted.

ONE DAY I overheard Scott saying that he estimated that the termites in a mound had approximately the computing power of a dog. His thinking was as follows: While a golden retriever has 627 million neurons, a termite might have about two hundred thousand. But as the termites glom together by the millions in their mound, their collective brain becomes something to reckon with. "A fair bit of calculating power," he said in a cerebral tone. And, with one of his trademark Socratic chess moves: "But you have to ask, What is a brain?"

Dogs think a lot, especially when you have something delicious in your pocket: I know the dog thinks because it thinks about me! Termites, though, are indifferent. I looked up from my notes and saw a large flock of birds wheeling, spreading and coalescing, creating a model of the landscape below among themselves in the air. Were they a brain?

One evening a few days later I asked Scott again about termites and cognition. We were sitting outside, looking across a field with a few soft red mounds still glowing in grass that had gone blue-gray in the deep dusk. Scott was writing a new book on homeostasis. "Homeostasis implies cognition," he said. Because the termites know where they, their zone of control, and their mound end in the landscape, they have a kind of whole awareness of where they are and what they are. It's not a precise knowing, the way I know I'm sitting in Namibia, but a more emergent group awareness of the boundaries between themselves and other—whether that is other dirt or another mound.

He said that the termites and fungi and microbes act together as a cognitive system. And he speculates that looking at these complex interdependencies, these social systems married with distinctive architecture and chemistry, could be more revealing than our current system of looking at individual organisms and their genes. "Rather than defining the individual genetically, it may be more productive to define them cognitively," he said.

Scott had come to think that the mounds themselves were a physical memory, with their mixture of shapes and smells and templates of gases, that allowed one generation of termites to pass their gains on to the next the way we hand down machines and books. This concept made them, in a sense, the architects of their own codes—in the balls of mud and spit of the mound—rather than robots who merely enacted the code written in their genes.

Did the termites have intentions, or at least intentionality? Scott cautioned against "a high flake factor," and then made his case. Cognition, he said, is a social phenomenon, whether it's happening as nerve cells interact with one another or within a crowd of termites. Together the termites have a sense of what their environment should be like, a sort of map of a proper world. This perfect mound has few breezes, has the perfect concentrations of carbon dioxide and humidity, has smooth edges and hard—not crumbly—walls. "What the termites collectively do is, they build a world to conform to that cognitive picture," he said, "and that is the intentionality of the swarm." And that would make the mound an expression by the collective dog-size brain of the termite's ideal world, a sort of utopia in dirt.

By that time it was so dark we couldn't see any mounds at all. Soon we were mostly surrounded by the sounds of rustling in the undergrowth, insects chirping, birds cooing, and some baboons calling in the distance. Mixed in with the communication of the creatures were the scents the flowers and grasses made to signal bugs and birds. It was not a peaceful night on the porch, but a dense and overwhelming experience—like listening to twelve foreign-language radio stations at once. There was much more information than I could hear or smell. Underneath me the termites were probably clicking while the microbes in my intestines sent chemical signals. Was I part of this complex signaling without my knowledge? Undoubtedly, but I don't know myself as an organism—and whatever other consciousnesses or cognitions I may encompass—only as that illusory unitary mind.

If termite mounds have cognition, and so (at some level) do groups of bacteria, zebrafish, and elephants, then we humans may not be so special. We could be obliviously hauling several pounds of mind through a world thick with the stuff.

I'VE READ AND REREAD *The Soul of the White Ant* many times. With each reading there are new things to admire as well as new frustrations: Marais's superorganism is part close observation, part riddle, part metaphor for the universe, and of course crackpot yarn. After the fifth reading or so I wondered whether that Belgian Nobel Prize winner Maurice Maeterlinck had really plagiarized Marais. I just didn't see how anyone could possibly adopt the whacked persona that had written *The Soul of the White Ant*.

I found Maeterlinck's *The Life of the White Ant* and read it in an afternoon. It was immediately obvious that the reason no one reads Maeterlinck's book anymore is that there are few big bizarre unwieldy ideas, only earnest facts, which were long ago surpassed by better facts. If he did plagiarize Marais, Maeterlinck did not take the good stuff.

Marais didn't write a termite science book, but a book about how humans could understand termites—as a bug, a body, a soul, a force on the landscape. Looking at termites this way changed how I see the world, science, the future, and myself—but it didn't waste my time on soon-to-spoil facts.

When I started following termites and scientists over the long term, I expected to watch scientific paradigms fall. This has in fact happened. The way we understand competition and altruism in social insects has shifted over the past decade, though the details remain contended. We have a new understanding of how termites and their gut microbes evolved together. And the development of giant protists like *Mixotricha paradoxa* and *Trichonympha* has come under new scrutiny. Within the termite mound, ideas such as stigmergy and pheromones as drivers of construction have been challenged and revised. Thanks to Kirstin's tracker, the whole idea of termites as uniform factory drudges has shifted spectacularly. Meanwhile, our understanding of spatial organization and ecology and our ability to model how they work has taken great leaps.

In the end, it wasn't the falling paradigms—or the fights and confusion they sometimes cause—that were the most interesting. These changes of perspective are the normal path of science. What I really enjoyed was seeing where juicy questions lead over the course of years of inquiry. And while mysteries are frustrating—especially if you're trying to reengineer an *E. coli* cell—they are also fascinating. What we don't know, and where we've failed, is just as interesting as success, however it's defined.

I was surprised by how many different versions of science—even of biology—there are. As the field struggles to become more predictive, or at least more quantitative, it is changing. Researchers like Scott, driven by questions and obsessions in the field, are being joined by consortia of researchers who cross disciplines, the way Rob and Corina's team pooled their knowledge of ecology and biological math. As others—the geneticists, the synthetic biologists, and even the roboticists—try to find meanings in vast quantities of data, they are running up against the limits to abstracting life's complexity. And the burden of trying to solve society's problems is also changing how biologists see their own work. After centuries of human efforts to analyze life, biology itself is evolving in divergent ways.

But all of this is really just a warm-up. If Carnot does show up, our Beckett play will be over and we'll be in a new story that hasn't been written yet. And even if we never get a unified theory, just lots and lots of new insights into the rules that drive living things, there will be changes in science, technology, the future of humans and of the planet. The arrival of these new ideas—and their implications for our place in the universe—will not only change our minds, they will change our idea of what a mind is.

In an earlier age of discovery, Descartes used his graphs of cause and effect, his metaphors about automata, and his habits of thought and analysis to give us a way to talk and think about the world. The task— for scientists and the rest of us—is to find a new common language for what we understand and also for what we don't.

AFTER FOLLOWING TERMITES for so long, I have to report that they've trailed me to Maine. Right now they've supposedly been found twenty miles south of where I sit, steadily munching northward, year by year, as the Gulf of Maine warms. Every year from now on I'll live in a place that I recognize less.

Soon some termites won't recognize themselves. In southern Florida the human process of urbanization has led to the spread of two invasive termites (*Coptotermes formosans* and *C. gestroi*). But climate change has made the timing of the two species' nuptial flights sync up. Recently, males of one species started preferring females of the other species to those of their own. Now the two species have begun to hybridize, forming colonies that grow at twice the speed of either of the originals, with individuals that researchers describe as potential "super-termites." While humans have been trying to engineer termite traits, termites have been reengineering themselves to take advantage of human behavior.

The termite species that is most likely to succeed is the ancient *Mastotermes darwiniensis*, which is expected to double its range by 2070. Though these cockroachy termites are currently confined to Australia, with the warming climate humans are likely to carry them to parts of southern Asia, across southern India, and to a wide swath of Africa,

South America, and the southern United States and parts of Mexico. They have already made it to Papua New Guinea, where they immediately attacked forty-two species of trees. I've been told that because it's too dry for a proper nuptial flight there, some of the queens have begun breeding larger colonies through parthenogenesis. Twelve of the thirteen most invasive termite species are likely to spread, meaning you'll soon have new neighbors, too.

If there is a termite apocalypse of some sort, it'll be fair to say that humans started it. But the relationship between termites and humans isn't really that straightforward. In some places, termites disappear when development shows up. For example, when woodlands in Sumatra are cleared for rubber plantations that are eventually turned into cassava gardens, the number of termite species found on the land decreases from thirty-four to one. Termites have a lot of tools—both genetic and behavioral—for surviving in rough times, but not all of them will. The species that survive may be those who are best suited to us.

Insects in general appear to be exceptionally endangered. A group of naturalists in Germany who've been measuring flying bugs for decades say that since 1989, the mass of insects they collect in nature reserves has fallen by 80 percent. Though data is limited, one global index measured a 45 percent decline in insect populations over the last forty years. But even this may be a miscount: only 1 million of the 5 million estimated insect species on Earth are named. They may be gone before we even know them.

The termites know, at one level or another, quite a bit about us. Over millennia, they have painstakingly aided in the construction of what we consider reality—fertile fields; hollow trees for our fires; geckos; elephants. But will they always?

NOTES

1. A TERMITE SAFARI

To get a general background sense of how bugs have evolved, David Grimaldi and Michael S. Engle's *Evolution of the Insects* (Cambridge, U.K.: Cambridge University Press, 2005) is wonderful: richly illustrated and clearly written. It's full of bizarre and fascinating tidbits of information that left me with the sense that bugs are the center of the world.

While doing research for this book, I used research by and interviews with many entomologists and termite researchers. Though I only quote some of them by name, their dogged investigations into the termites gave me some of the basic background and hairy details of this story: Rudi Scheffrahn, David Bignell, Rebeca Rosengaus, Paul Eggleton, Christine Nalepa, Theo Evans, Michael Haverty, Judith Korb, Susan Jones, and Vernard Lewis, to name a few.

3 *a story for* The Atlantic: Margonelli, "Gut Reactions," *The Atlantic*, September 2008, www.theatlantic.com/magazine/archive/2008/09/gut-reactions /306946/.

4 *Phil and thirty-eight other scientists who sequenced*: Falk Warnecke, Peter Luginbühl, Natalia Ivanova et al., "Metagenomic and Functional Analysis of Hindgut Microbiota of a Wood-Feeding Higher Termite," *Nature* 450 (2007): 560–65.

4 *last survey of termites in the area, published in 1934*: Charles A. Kofoid, Sol Felty Light, A. C. Horner et al., eds., *Termites and Termite Control: A Report to the Termite Investigations Committee* (Berkeley: University of California Press, 1934).

5 *Only twenty-eight out of twenty-eight hundred species*: Theodore A. Evans, Brian T. Forschler, and J. Kenneth Grace, "Biology of Invasive Termites: A Worldwide Review," *Annual Review of Entomology* 58 (2013): 455–74.

5 *Rudi believes*: Rudolf H. Scheffrahn, Jan Krecek, Renato Ripa, Paola Luppichini, "Endemic Origin and Vast Anthropogenic Dispersal of the West Indian Drywood Termite," *Biological Invasions* 11 (2009): 796.

6 *In 2007 a paper was published*: Daegan Inward, George Beccaloni, and Paul Eggleton, "Death of an Order: A Comprehensive Molecular Phylogenetic Study Confirms That Termites Are Eusocial Cockroaches," *Biology Letters* 3 (2007): 331–35.

6 *became obvious with DNA sequencing*: David Grimaldi and Michael S. Engel, *Evolution of the Insects* (Cambridge, U.K.: Cambridge University Press, 2006), 237.

10 *Between 2000 and 2013, 6,373 papers*: David E. Bignell, "The Role of Symbionts in the Evolution of Termites and Their Rise to Ecological Dominance in the Tropics," in *The Mechanistic Benefits of Microbial Symbionts*, ed. Christon J. Hurst (Basel: Springer International, 2016), 137.

10 *somewhere between $1.5 and $20 billion*: Anne Nagro, "A-maze-ing Journey," *Pest Control Technology*, February 26, 2015, www.pctonline.com/article/pct0215-annual-termite-damage-quest/.

10 *outweigh us ten to one*: Bignell, "Role of Symbionts," 136.

10 *oceans would become toxic*: Grimaldi and Engel, *Evolution of the Insects*, 5–6.

11 *the Hoover Dam altered*: U.S. Department of the Interior, Bureau of Reclamation: "The Colorado River and Hoover Dam: Facts and Figures," www.usbr.gov/lc/region/pao/faq.html.

12 *the thing about stonewashed jeans*: Rita Araújo, Margarida Casal, and Artur Cavaco-Paulo, "Application of Enzymes for Textile Fibres Processing," *Biocatalysis and Biotransformation* 26 (2008): 332–49.

2. RIDDLES IN THE DIRT

There are several editions of Eugène Marais's book available in the United States, and one of them has been abridged, which really diminishes its charms. Try to find an older copy. I kept rereading *The Soul of the White Ant* during the years I was working on this book, intrigued by its stories within stories and its slippery relationship to truth. I liked Doris Lessing's description of his work:

"He offers a vision of Nature as a whole, whose parts obey different time laws, move in affinities and linkages we could learn to see, parts making wholes on their own levels, but seen by our divisive brains as a multitude of individualities, a flock of birds, a species of plant or beast—man. We are just at the start of an understanding of the heavens as a web of interlocking clocks, all differently set: an understanding that is not intellectual but woven into experience." ("Ant's Eye View: A Review of *The Soul of the White Ant* by Eugène Marais," in *A Small Personal Voice: Essays, Reviews, Interviews* [New York: Vintage, 1975], 146.) But you can't separate the story he tells from his complicated biography (well covered in *The Dark Stream*, cited below) or his symbolic significance as a leading Afrikaans poet.

15 *on assignment from* National Geographic: Margonelli, "Collective Mind in the Mound: How Do Termites Build Their Huge Structures?," *National Geographic* Online, August 1, 2014, https://news.nationalgeographic.com /news/2014/07/140731-termites-mounds-insects-entomology-science/.

16 *constructing shelters on Mars*: Clive Cookson, "Termites Teach Scientists to Build for the Future," *Financial Times*, September 25, 2004, 5.

17 *they show up on weather radar*: Willborn P. Nobles III, "Termite, Insect Swarms Across New Orleans Picked Up on Meteorologists' Radar," *The Times-Picayune*, May 30, 2016, www.nola.com/environment/index.ssf/2016 /05/meteorologists_see_termite_ins.html.

17 *catching and eating these termites*: Julie J. Lesnik, "Termites in the Hominin Diet: A Meta-Analysis of Termite Genera, Species and Castes as a Dietary Supplement for South African Robust Australopithecines," *Journal of Human Evolution* 71 (2014): 94–104.

18 *"The first thing she does"*: Eugène N. Marais, *The Soul of the White Ant*, trans. Winifred de Kok (New York: Dodd, Mead, 1937; repr. Penguin, 1973).

19 *Marais was a journalist*: Leon Rousseau, *The Dark Stream: The Story of Eugène N. Marais* (Johannesburg: Jonathan Ball, 1982), 26, 52.

20 *analyzed the queen pheromone*: Kenji Matsuura, "Multifunctional Queen Pheromone and Maintenance of Reproductive Harmony in Termite Colonies," *Journal of Chemical Ecology* 38 (2012): 746–54.

20 *researchers Christine Nalepa, Theo Evans, and Michael Lenz*: Christine A. Nalepa, Theodore A. Evans, and Michael Lenz, "Antennal Cropping During Colony Foundation in Termites," *Zookeys* 148 (2011): 185–96.

21 *some morph into soldiers*: Hajime Yaguchi, Takaya Inoue, Ken Sasaki, and Kiyoto Maekawa, "Dopamine Regulates Termite Soldier Differentiation Through Trophallactic Behaviours," *Royal Society Open Science* 3 (2016): 150574.

21n *Soldiers appear to return*: Shulin He, Paul R. Johnston, Benno Kuropka, et al., "Termite Soldiers Contribute to Social Immunity by Synthesizing Potent Oral Secretions, *Insect Molecular Biology*, April 17, 2018.

22 *multiple queens*: Tamara R. Hartke and Rebeca B. Rosengaus, "Costs of Pleometrosis in a Polygamous Termite," *Proceedings of the Royal Society B: Biological Sciences* 280 (2013): 20122563.

22 *founded by two male termites*: Nobuaki Mizumoto, Toshihisa Yashiro, and Kenji Matsuura, "Male Same-Sex Pairing as an Adaptive Strategy for Future Reproduction in Termites," *Animal Behaviour* 119 (2016): 179–87.

22 *the entomologist William Wheeler*: William Morton Wheeler, "The Ant-Colony as an Organism," *Journal of Morphology* 22 (1911): 307–25.

23 *the dictatorship in* Brave New World: Charlotte Sleigh, *Six Legs Better: A Cultural History of Myrmecology* (Baltimore: Johns Hopkins University Press, 2007), 64.

23n *Wheeler supported Herbert Hoover's*: Sleigh, *Six Lives Better*, 82–85.

24 *Formosan subterranean termite was transplanted*: Matthew T. Messenger, Nan-Yao Su, and Rudolf H. Scheffrahn, "Current Distribution of the Formosan Subterranean Termite and Other Termite Species (Isoptera: Rhinotermitidae, Kalotermitidae) in Louisiana," *Florida Entomologist* 85 (2002): 580–87.

24 *the superorganism became "a pariah"*: Abraham H. Gibson, "Edward O. Wilson and the Organicist Tradition," *Journal of the History of Biology* 46 (2012): 599–630.

25 *Wilson himself had recently written*: Bert Hölldobler and Edward O. Wilson, *The Superorganism: The Beauty, Elegance, and Strangeness of Insect Societies* (New York: W. W. Norton, 2009).

25 *One day in 1999*: Richard Dawkins and Steven Pinker met for the Guardian-Dillons Debate at the Westminster Central Hall in London on February 10, 1999. The debate was chaired by Tim Radford. The transcript is archived at www.edge.org/conversation/richard_dawkins-steven_pinker-is-science-killing-the-soul.

3. AN INCONVENIENT INSECT

When he first started working on termite mounds in Namibia, Scott worked and lived out of a van. Then he got a grant from the nonprofit Earthwatch, which provided him with volunteers who wanted a sort of working vacation. Later he became friends with Eugene Marais, who helped him get access to Omatjenne, which was a government-operated farm doing research on cattle raising. In 2013 Scott moved his research to another location owned by the Cheetah Conservation Fund. Scott's recent thinking can be found in *Purpose and Desire: What Makes Something "Alive" and Why Modern Darwinism Has Failed to Explain It* (New York, HarperCollins, 2017).

26 *out of town to Omatjenne*: Scott has written about the farm here: www.esf.edu/efb/turner/termitePages/Omatjenne.html.

28 *19 degrees from north*: J. Scott Turner, "Architecture and Morphogenesis in the Mound of *Macrotermes michaelseni* (Sjöstedt) (Isoptera: Termitidae, Macrotermitinae) in Northern Namibia," *Cimbebasia* 16 (2000): 143–75. For more-general information: www.esf.edu/efb/turner/termitePages /termiteStruct.html.

29 *at a nursing school*: J. Scott Turner, *The Tinkerer's Accomplice: How Design Emerges from Life Itself* (Cambridge, Mass.: Harvard University Press, 2007), 15.

30 *famously described as working like chimneys*: Martin Lüscher, "Die Lufter-neuerung im Nest der Termite *Macrotermes natalensis* (Haviland)," *Insectes Sociaux* 3 (1956): 273–76.

30 *years of experiments*: J. Scott Turner, *The Extended Organism: The Physiology of Animal-Built Structures* (Cambridge, Mass.: Harvard University Press, 2000), 185–200.

31 *Rupert had recently left*: Daniel Pimlott, "Natural-Born Housebuilders," *Financial Times*, July 17, 2009.

4. INTO THE MOUND

Charlotte Sleigh's *Six Legs Better* (cited above) and Diane M. Rodgers's book *Debugging the Link Between Social Theory and Social Insects* (cited below) both unpack the fascinating and disturbing story of how Europeans and Americans have conflated human society and insect societies. Rodgers looks at science and sociology, while Sleigh looks more at popular culture. You can see Henry Smeathman's watercolors of termite mounds at the Royal Society (https://pictures.royalsociety.org/image-rs-8497). In early drafts of the book I attempted to understand Smeathman's career, which involved being both an abolitionist and a slave trader. But I gave up—it deserves its own book. J.F.M. Clark's *Bugs and the Victorians* (New Haven: Yale University Press, 2009) has an early chapter that delves into Smeathman's illustrations and life.

35 *the British biologist J.B.S. Haldane*: Lee Alan Dugatkin, "Inclusive Fitness Theory from Darwin to Hamilton," *Genetics* 176 (2007): 1375–80.

36 *In 1965 the entomologist E. O. Wilson*: Edward O. Wilson, *Naturalist* (Washington, D.C.: Island Press, 2006), 320.

36 *In 2010 . . . he published a paper*: Martin A. Nowak, Corina E. Tarnita, and Edward O. Wilson, "The Evolution of Eusociality," *Nature* 466 (2010): 1057–62.

36 *In response, more than one hundred scientists wrote*: Patrick Abbot, Jun Abe, John Alcock et al., "Inclusive Fitness Theory and Eusociality," *Nature* 471 (2011): E1–E4.

38 *ancient San people made thousands of rock drawings*: Siyakha Mguni, "Iconography of Termites' Nests and Termites: Symbolic Nuances of Formlings in Southern African San Rock Art," *Cambridge Archaeological Journal* 16 (2006): 53–71.

40 *a fungus that they have been cohabiting with*: Duur K. Aanen, Henrik H. de Fine Licht, Alfons J. M. Debets et al., "High Symbiont Relatedness Stabilizes Mutualistic Cooperation in Fungus-Growing Termites," *Science* 326 (2009): 1103–1106.

41 *eight times bigger*: Scott based his estimates on his own research and several papers, among them this one: Jo P.E.C. Darlington, Patrick R. Zimmerman, James Greenberg, et al., "Production of Metabolic Gases by Nests of the Termite *Macrotermes jeanneli* in Kenya," *Journal of Tropical Ecology* 13 (1997): 491–510.

41n *Fungi are capable*: Kenji Matsuura, Chihiro Tanaka, and T. Nishida, "Symbiosis of a Termite and a Sclerotium-Forming Fungus: Sclerotia Mimic Termite Eggs," *Ecological Research* 15 (2000): 405–14.

42 *a rind of crystalline material*: X. Liu, H. Curtis Monger, and Walter George Whitford, "Calcium Carbonate in Termite Galleries—Biomineralization or Upward Transport?," *Biogeochemistry* 82 (2007): 241–50.

43 *Some species of termite queens*: Toshihisa Yashiro and Kenji Matsuura, "Termite Queens Close the Sperm Gates of Eggs to Switch from Sexual to Asexual Reproduction," *Proceedings of the National Academy of Sciences* 111 (2014): 17212–17.

44 *It wasn't until the 1670s*: Diane M. Rodgers, *Debugging the Link Between Social Theory and Social Insects* (Baton Rouge: Louisiana State University Press, 2008), 44.

44 *Smeathman gave to the Royal Society in 1781*: Henry Smeathman, "Some Account of the Termites, Which Are Found in Africa and Other Hot Climates. In a Letter from Mr. Henry Smeathman, of Clement's Inn, to Sir Joseph Banks, Bart., P.R.S," *Philosophical Transactions of the Royal Society of London* 71 (1781): 139–92; 139, 145.

45 *darker ants as "slaves"*: Rodgers, *Debugging the Link*, 132–34; 176–77.

45 *the Russian zoologist Pyotr Alexeyevich Kropotkin*: James T. Costa, "Scale Models? What Insect Societies Teach Us About Ourselves," *Proceedings of the American Philosophical Society* 146 (2002): 170–80.

45 *1919 comic speech*: William Morton Wheeler, "The Termitodoxa, or Biology and Society," *The Scientific Monthly* 10 (1920): 113–24; 117, 118, 119, 121–23.

45n *Gilman saw women's current low status*: Diane M. Rodgers, *Debugging the Link Between Social Theory and Social Insects*, 159.

46 *the ant scientist Deborah Gordon*: Deborah M. Gordon, "The Organization of Work in Social Insect Colonies," *Nature* 380 (1996): 121.

46 *"the firing patterns of neurons in the brain"*: Janet Basu, "Down on the (Ant) Farm: Insects Yield Clues to How Brain Cells Work," *Stanford Today*, May–June 1996, web.stanford.edu/dept/news/stanfordtoday/ed/9605/9605ants.html.

5. COMPLEXITY IS THE ESSENCE

The literature on stigmergy and swarm intelligence is vast. I read the popular science versions with skepticism because they tend to treat stigmergy as a "natural" solution to very human problems like routing phone calls. Recent research raises the question of whether stigmergy is less a natural process than an artifact of the way humans see termites. Likewise, swarm intelligence—often touted as a solution to problems—is also an interesting way to think about thinking. A good place to start is the essay "Swarm Smarts," by Eric Bonabeau and Guy Théraulaz, originally published in the March 2000 issue of *Scientific American.*

48 *The more he thought about social insects:* Turner, *Extended Organism,* 130–32.
48 *A French termite researcher named Pierre-Paul Grassé:* Guy Théraulaz and Eric Bonabeau, "A Brief History of Stigmergy," *Artificial Life* 5 (1999): 97–116.
49 *simulations featuring virtual termites:* Mitchel Resnick, *Turtles, Termites, and Traffic Jams: Explorations in Massively Parallel Microworlds* (Cambridge, Mass.: MIT Press, 1994).
49 *sometimes called swarm intelligence:* Eric Bonabeau and Guy Théraulaz, "Swarm Smarts," *Scientific American* 282 (March 2000): 72–79.
49 *Eric Bonabeau, an early leader in the field:* Derrick Story, "Swarm Intelligence: An Interview with Eric Bonabeau," *OpenP2P,* February 21, 2003, www.openp2p.com/pub/a/p2p/2003/02/21/bonabeau.html.

6. BECAUSE THEY ARE SO SWEET

How can something think without a head, or even a central nervous system? Recent work has shown that types of cognition (which is not to be confused with self-awareness) happen at many levels, from signaling within cells, to communication within slime mold communities, to brain-like command centers in the roots of trees, to memory in bone cells. For a mind-blowing tour of the subject, read František Baluška and Michael Levin, "On Having No Head: Cognition Throughout Biological Systems," *Frontiers in Psychology* 7 (2016): 902.

54 *"Our ancestors did not start society because":* Wheeler, "Termitodoxa," 115.
56 *an oily hydrocarbon molecule:* Michael I. Haverty, R. Nelson Woodrow, Lori J. Nelson, and J. Kenneth Grace, "Identification of Termite Species by the Hydrocarbons in Their Feces," *Journal of Chemical Ecology* 31 (2005): 2119–51.
56 *feces spread on the walls that repel invading fungi:* Thomas Chouvenc, Caroline A. Efstathion, Monica L. Elliott, and Nan-Yao Su, "Extended Disease Resistance Emerging from the Faecal Nest of a Subterranean Termite," *Proceedings of the Royal Society B: Biological Sciences* 280 (2013): 20131885.

56 *These tricks, or cognitive traps*: J. Scott Turner, "Termites as Models of Swarm Cognition," *Swarm Intelligence* 5 (2011): 19–43.

57 *The U.S. Army was paying for these experiments*: Ibid.

58 *The biggest mushrooms were worth twenty dollars*: N'Golo Koné, Kolo Yéo, Souleymane Konaté, and Karl Eduard Linsenmair, "Socio-economical Aspects of the Exploitation of Termitomyces Fruit Bodies in Central and Southern Cote d'Ivoire: Raising Awareness for Their Sustainable Use," *Journal of Applied Biosciences* 70 (2013): 5580–90.

59 *beat a lizard until it confessed to being a crocodile*: Jim Hooper, *Koevoet!: Experiencing South Africa's Deadly Bush War* (Solihull, West Midlands, U.K.: Helion & Company and Rugby, Warwickshire: GG Books UK, 2013), 63.

59 *the first genocide of the twentieth century*: Marion Wallace with John Kinahan, *A History of Namibia: From Earliest Times to 1990* (New York: Oxford University Press, 2013), 177–79.

7. A BLACK BOX WITH SIX LEGS

For a deeper understanding of the concepts underlying the roboticists' work, read *Complexity: A Guided Tour*, by Melanie Mitchell (Oxford, U.K.: Oxford University Press, 2009). A professor of computer science who is affiliated with the Santa Fe Institute, she explains some of the big intellectual ideas behind how small and simple things—like termites and neurons—can create complex structures like mounds and brains. If you still want more, you can take a "Complexity Explorer" class online at the Santa Fe Institute.

61 *"Programmable Second Skin"*: NSF Award # 0932015, "Programmable Second Skin to Re-educate Injured Nervous Systems," National Science Foundation, September 2009, www.nsf.gov/awardsearch/showAward?AWD_ID =0932015.

61 *"RoboBees: A Convergence"*: NSF Award # 0926148, "Collaborative Research: RoboBees: A Convergence of Body, Brain and Colony," National Science Foundation, August 2009, www.nsf.gov/awardsearch/showAward ?AWD_ID=0926148.

62 *audience member asked*: Matt Welsh, "RoboBees: An Autonomous Colony of Robotic Pollinators" (presentation at 2010 USENIX Annual Technical Conference), published online August 18, 2010, https://www.you tube.com/watch?v=-bLTiYkYyVc. The discussion appears at the 49th minute.

64 *programming the lab's termite-based robots*: Justin Werfel, "Anthills Built to Order: Automating Construction with Artificial Swarms" (PhD dissertation, MIT, 2006), http://hdl.handle.net/1721.1/33791.

68 *Descartes who described*: Desmond M. Clarke, *Descartes: A Biography* (Cambridge, U.K.: Cambridge University Press, 2006; repr. 2012), 321–24.

68 *a naked mechanical Neptune*: Jessica Riskin, "Machines in the Garden," *Republics of Letters* 1, no. 2 (April 30, 2010), http://arcade.stanford.edu/rofl/machines-garden.

69 *through Descartes's eyes*: Stephen Gaukroger, *Descartes: An Intellectual Biography* (Oxford, U.K.: Oxford University Press, 1995; repr. 2003), 63, 149, 393.

8. WAITING FOR CARNOT

How does technology change what we are capable of as well as our moral imagination? Reading Wolfgang Schivelbusch's *The Railway Journey: The Industrialization of Time and Space in the Nineteenth Century* (Oakland: University of California Press, 1986) showed how riding trains changed social relations, the way we see land and time, and our sense of what constitutes a disaster. The process reminded me of the Internet. Later I found John Ellis's *Social History of the Machine Gun* (cited below), which explained why the machine gun was used in Africa as well as why Europeans were unprepared for its moral disruption. The story of how technology and ideology interacted to create several levels of disaster, first in Africa and Tibet, and then in Europe, seems key for imagining how technology will affect our own future.

76 *Over the course of the daylong battle*: M. Wallace, *History of Namibia*, 155–182, with particular details on pages 162–63.

76 *"rivers of blood and money"*: George Steinmetz, "The First Genocide of the 20th Century and Its Postcolonial Afterlives: Germany and the Namibian Ovaherero," *Journal of the International Institute* 12, no. 2 (2005), 64, http://hdl.handle.net/2027/spo.4750978.0012.201.

77 *Repatriation of the skulls*: David Knight, "Skulls of Colonial Victims Returned to Namibia," *Der Spiegel*, September 27, 2011, www.spiegel.de/international/germany/there-was-injustice-skulls-of-colonial-victims-returned-to-namibia-a-788601.html.

77 *sentimentality and excuses for genocide*: W. Winwood Reade, *Savage Africa: Being the Narrative of a Tour in Equatorial, Southwestern, and Northwestern Africa* (New York: Harper and Brothers, 1864), 452, https://catalog.hathitrust.org/Record/100209352. I learned of Reade's writings in David Olusoga and Casper W. Erichson, *The Kaiser's Holocaust: Germany's Forgotten Genocide and the Colonial Roots of Nazism* (London: Faber and Faber, 2011).

77 *The Social History of the Machine Gun*: John Ellis, *The Social History of the Machine Gun* (New York: Pantheon, 1975; repr. Baltimore: Johns Hopkins University Press, 1986), 17.

9. THE SECOND TERMITE SAFARI

Though microbes are now understood and described by their genes, for most of the history of microbiology, microbes were described by their shapes and behaviors. In 1964 the physicist-turned-microbiologist Carl Woese started comparing the genes on fragments of the single-stranded RNA for the ribosomes (the tiny molecular machines that build proteins) from different microbes. He used an enzyme to cut apart the RNA strands at the nucleotide G, labeled them with radioactive phosphorus, and pulled them through electrically charged gels to sort the molecules by size. Sometimes these gels were cooled by kerosene—a method forbidden in most labs today. Woese used X-ray film to capture the cascade of wobbly dots made by the molecules of DNA—some clustered and others far apart—as though the spots were escaping from a cartoon leopard. By comparing these prints, Woese was able to see how the microbes were related to one another and where they differed. Each evening as he headed home he'd say to himself, "Woese, you have destroyed your mind again today." By 1976, the team had X-rays and sequences for between thirty and sixty microbes.

He was lucky: the 16s subunit was a sort of genetic clock with an hour hand and a minute hand of evolutionary changes that were both slow and fast—which made it a kind of living fossil, reaching back to the beginnings of biological time but ticking off much smaller changes along the way. With his data, Woese was able to construct a phylogenetic tree—or tree of life—that showed the enormous evolutionary diversity of bacteria, the eukaryotes (cells with nuclei that include microbes, fungi, plants, animals, and you), as well as archaea, a newly identified kingdom of methane digesters he and a collaborator discovered in a cow's stomach. Though his work is now celebrated, and whole fields—including metagenomics—are based upon it, at the time Woese struggled for recognition and reportedly consoled himself by reading Thomas S. Kuhn's *The Structure of Scientific Revolutions* (Chicago: University of Chicago Press, 1962; 50th Anniversary edition, 2012). For more on Woese, see Norman R. Pace, Jan Sapp, and Nigel Goldenfeld, "Phylogeny and Beyond: Scientific, Historical, and Conceptual Significance of the First Tree of Life," *Proceedings of the National Academy of Sciences* 109 (2012): 1011–18; and David W. Wolfe, "The Empire Underground," *The Wilson Quarterly* 25 (2001): 18–27.

83 *termite gut as a veritable factory*: Andreas Brune, "Symbiotic Digestion of Lignocellulose in Termite Guts," *Nature Reviews Microbiology* 12 (2014): 168–81.

85 *When Phil and thirty-eight other researchers*: Falk Warnecke, Peter Luginbühl, Natalia Ivanova et al., "Metagenomic and Functional Analysis."

85 *enzymes that might work to digest wood*: Luen-Luen Li, Sean R. McCorkle, Sebastien Monchy et al., "Bioprospecting Metagenomes: Glycosyl Hydrolases for Converting Biomass," *Biotechnology for Biofuels* 2 (2009): 10.

85 *Press releases suggested*: "DOE JGI Plumbs Termite Guts to Yield Novel Enzymes for Better Biofuel Production," JGI News Release, November 21, 2007, https://jgi.doe.gov/news_11_21_07/.

85 *focused on the grassoline problem*: I wrote on the beginnings of the cluster of laboratories and the difficulties of monitoring their work. Margonelli,

"Start-Up U," *California Magazine*, September–October 2007, https://alumni
.berkeley.edu/california-magazine/september-october-2007-green-tech
/start-u.

87 *fourteen hundred different species of bacteria*: Brittany F. Peterson and
Michael E. Scharf, "Lower Termite Associations with Microbes: Synergy,
Protection, and Interplay," *Frontiers in Microbiology* 7 (2016): 422.

88 Trichonympha, *the most common*: Kevin J. Carpenter, Lawrence Chow, and
Patrick J. Keeling, "Morphology, Phylogeny, and Diversity of *Trichonympha* (Parabasalia: Hypermastigida) of the Wood-Feeding Cockroach *Cryptocercus Punctulatus*," *Journal of Eukaryotic Microbiology* 56 (2009): 305–13.

88 Trichonympha *owes its name*: Joseph Leidy, "On Intestinal Parasites of
Termes flavipes," *Proceedings of the Academy of Natural Sciences of Philadelphia* 29 (1877): 146–49; I first saw mention of this on Patrick Keeling's
website: http://www3.botany.ubc.ca/keeling/invisible/index.html.

88 *DNA from whole communities*: David J. Lane, Bernadette Pace, Gary J. Olsen et al., "Rapid Determination of 16S Ribosomal RNA Sequences for
Phylogenetic Analyses," *Proceedings of the National Academy of Sciences* 82
(1985): 6955–59.

88 *studying microorganisms*: Philip Hugenholtz, Christian Pitulle, Karen L.
Hershberger, and Norman R. Pace, "Novel Division Level Bacterial Diversity in a Yellowstone Hot Spring," *Journal of Bacteriology* 180 (1998): 366–76.

89 *Jill Banfield's lab*: Gene W. Tyson, Jarrod Chapman, Philip Hugenholtz
et al., "Community Structure and Metabolism Through Reconstruction
of Microbial Genomes from the Environment," *Nature* 428 (2004): 37–43.

89 *By 2007, the startling ability of metagenomics*: Committee on Metagenomics
and Board on Life Sciences, National Research Council of the National
Academies, *The New Science of Metagenomics: Revealing the Secrets of Our
Microbial Planet* (Washington, D.C.: National Academies Press, 2007). Available from www.ncbi.nlm.nih.gov/books/NBK54011/.

89 *Ninety percent of the microbes*: Margaret McFall-Ngai, Michael G. Hadfield,
Thomas C. G. Bosch et al., "Animals in a Bacterial World, a New Imperative for the Life Sciences," *Proceedings of the National Academy of Sciences* 110
(2013): 3229–36.

90n *Consider pandas*: Maureen A. O'Malley, *Philosophy of Microbiology* (Cambridge, U.K.: Cambridge University Press, 2014), 151.

91n *That survey*: Alan Burdick, "The Old Men and the Sea," *New York Times
Magazine*, December 31, 2006, www.nytimes.com/2006/12/31/magazine
/31hedgpeth.html.

92 *In 1974 an ambitious*: W. L. Nutting, M. I. Haverty, and J. P. LaFage, "Physical and Chemical Alteration of Soil by Two Subterranean Termite
Species in Sonoran Desert Grassland," *Journal of Arid Environments*, 12
(1987): 233–39; 235.

92 *estimate in the 1980s*: P. R. Zimmerman, J. P. Greenberg, S. O. Wandiga,
and P. J. Crutzen, "Termites: A Potentially Large Source of Atmospheric

Methane, Carbon Dioxide, and Molecular Hydrogen," *Science* (November 5, 1982): 563–65.

92 *3–4 percent of methane emissions*: Adrian Ho, Hans Erens, Basile Bazirake Mujinya, et al., "Termites Facilitate Methane Oxidation and Shape the Methanotrophic Community," *Applied and Environmental Microbiology* 79 (December 2013).

10. LIFE IN THE FIREHOSE

When I first arrived at JGI, researchers were still using a painstaking process to sequence DNA called Sanger sequencing that involved handpicking colonies of bacteria from petri dishes. Signs on the wall warned of the dangers of repetitive stress. But within that year, the lab moved to using several new automated high-throughput sequencing machines. Originally I had a very long description of ways to sequence DNA that my editors wisely said did not belong in a termite book. But there is a fascinating book about the invention of PCR, or polymerase chain reaction—a process that amplifies small sequences of DNA so they can be sequenced. Written by the anthropologist Paul Rabinow, it's worth reading not only for the unlikely story of PCR's mix of discovery, accident, and exploitation, but also because it is an anthropologist's view of the scientists and institutions involved: *Making PCR: A Story of Biotechnology* (Chicago: University of Chicago Press, 1996).

96 *2009, their output had risen eightfold*: JGI web page: "DOE Metrics and Statistics," https://jgi.doe.gov/our-projects/statistics/.

97 *outstripped the discovery of stars*: Zachary D. Stephens, Skylar Y. Lee, Faraz Faghri et al., "Big Data: Astronomical or Genomical?," *PLoS Biology* 13 (2015): e1002195.

97 *cost of reading a genome*: NIH National Human Genome Research Institute, "Cost of Sequencing a Human Genome," updated July 6, 2016, https://www.genome.gov/sequencingcosts/.

98 *Coomassie blue*: Anirban Mitra, "Coomassie Blues: A Colonial Legacy (?) in Molecular Biology," *ClubSciWri*, May 11, 2017, http://www.sciwri.club/archives/3483.

98 *since before Rosalind Franklin met Watson and Crick*: Franklin's notebooks from 1951 and '52 are archived online by the Wellcome Library: https://wellcomelibrary.org/item/b19832059#?c=0&m=0&s=0&cv=0&z=0.4789%2C0.0524%2C2.0077%2C1.3904.

100 *researchers at the Department of Energy*: R. D. Perlack et al., "Biomass as Feedstock for a Bioenergy and Bioproducts Industry: The Technical Feasibility of a Billion-Ton Annual Supply," U.S. Department of Energy and U.S. Department of Agriculture, 2005.

100 *estimate was updated to 100 billion gallons*: "Bioenergy Frequently Asked Questions," Department of Energy, Office of Energy Efficiency and

Renewable Energy, http://www.energy.gov/eere/bioenergy/bioenergy
-frequently-asked-questions#biomass.

100 *emissions from driving by as much as 86 percent*: Michael Wang, May Wu, and
Hong Huo, "Life-Cycle Energy and Greenhouse Gas Emission Impacts of
Different Corn Ethanol Plant Types," *Environmental Research Letters* 2
(2007): 024001.

100 *The future of biology Keasling imagines*: Jay D. Keasling, "Manufacturing Molecules Through Metabolic Engineering," *Science* 330 (2011): 1355–58. Available in proceedings of Institute of Medicine (U.S.) Forum on Microbial
Threats, Washington, D.C., 2011: www.ncbi.nlm.nih.gov/books/NBK84444/.

103 *a paper on a strange repeating series*: Victor Kunin, Rotem Sorek, and Philip
Hugenholtz, "Evolutionary Conservation of Sequence and Secondary
Structures in CRISPR Repeats," *Genome Biology* 8 (2007): R61.

104 *In 2002 or so, Keasling*: Robert Sanders, "Launch of Antimalarial Drug a
Triumph for UC Berkeley, Synthetic Biology," *Berkeley News*, April 11,
2013, http://news.berkeley.edu/2013/04/11/launch-of-antimalarial-drug
-a-triumph-for-uc-berkeley-synthetic-biology/.

105 *anthropologist Gaymon Bennett*: Gaymon Bennett, "The Moral Economy of
Biotechnical Facility," *Journal of Responsible Innovation* 2 (2015): 128–32.

105 *Craig Venter announced that his lab had created a cell*: Eleonore Pauwels,
"Communication: Mind the Metaphor," *Nature* 500 (2013): 523–24.

105 *In 1998 he challenged*: Michael D. Lemonick, "J. Craig Venter. Gene Map-
per," *TIME*, December 25, 2000.

105 *In 2011 I saw him speak*: On October 18, 2011, he spoke at the New America
Foundation, where he was interviewed by Michael Crow. I took notes. The
event was also covered by David Biello for *Scientific American*: https://www
.scientificamerican.com/article/can-algae-feed-the-world-and-fuel-the
-planet/.

106 *In the summer of 2010*: Presidential Commission for the Study of Bioethical
Issues, *New Directions: The Ethics of Synthetic Biology and Emerging Technologies* (Washington, D.C.: Presidential Commission for the Study of Bioethical Issues, 2010).

11. JAZZ IN THE METAGENOME

Metagenomics has revealed how our microbial DNA exceeds the DNA in our
cells, and called into question who we—as individuals—are. For a good book
on how advances in microbiology are challenging philosophy, read Maureen
O'Malley's *Philosophy of Microbiology* (cited above). The book serves as an introduction to what microbiology is discovering, as well as some of the explicit
and implicit philosophical issues in the science.

108 *The group was going to compare*: Shaomei He, Natalia Ivanova, Edward
Kirton et al., "Comparative Metagenomic and Metatranscriptomic

Analysis of Hindgut Paunch Microbiota in Wood- and Dung-Feeding Higher Termites," *PLoS ONE* 8 (2013): e61126.

110 *a single spirochete*: Harold Kirby, "Protozoa in Termites of the Genus *Amitermes*," *Parasitology* 24 (October 1932): 289–304.

111 *The microbial philosophers*: John Dupré and Maureen O'Malley, "Metagenomics and Biological Ontology," *Studies in History and Philosophy of Science Part C: Biological and Biomedical Sciences* 38 (2007): 834–46.

111 *Nitrogen makes up 78 percent*: Scott Fields, "Global Nitrogen: Cycling Out of Control," *Environmental Health Perspectives* 112 (2004): A556–63.

114 *Consider* Crocosphaera: Jonathan P. Zehr, Shellie R. Bench, Elizabeth A. Mondragon et al., "Low Genomic Diversity in Tropical Oceanic N_2-Fixing Cyanobacteria," *Proceedings of the National Academy of Sciences* 104 (2007): 17807–12.

115 *comparing microbes in the guts of sixty-six*: Nurdyana Abdul Rahman, Donovan H. Parks, Dana L. Willner et al., "A Molecular Survey of Australian and North American Termite Genera Indicates That Vertical Inheritance Is the Primary Force Shaping Termite Gut Microbiomes," *Microbiome* 3 (2015): 5.

116 *More-recent work suggests*: Stephan J. Ott, Georg H. Waetzig, Ateequr Rehman et al., "Efficacy of Sterile Fecal Filtrate Transfer for Treating Patients with *Clostridium difficile* Infection," *Gastroenterology* 152 (2017): 799–811.e7.

12. BURNING VERY SLOWLY

At the beginning of the 2000s, the attempt to engineer life was cross-disciplinary. Radhika Nagpal, who later became a roboticist, collaborated with Drew Endy, who became a leading theorist of synthetic biology. Revisiting those early days, which weren't that long ago, shows how ideas moved freely back and forth between different fields. In Endy's 2005 article, "Foundations for Engineering Biology" (*Nature* 438: 449–53), he questions whether our failure to engineer biology is the result of biology's inherent rule-lessness or because we've failed to apply engineering or other systems of abstraction hard enough. Five years later, this question still haunted the field, as Roberta Kwok wrote in "Five Hard Truths for Synthetic Biology" (*Nature* 463 [2010]: 288–90).

118 *flowchart of how the termite's gut breaks down wood*: A version of this drawing can be found on page 5 of Shaomei He et al.'s "Comparative Metagenomic and Metatranscriptomic Analysis of Hindgut Paunch Microbiota in Wood- and Dung-Feeding Higher Termites."

119 *classic problems from biology*: Héctor García Martín and Nigel Goldenfeld, "On the Origin and Robustness of Power-Law Species–Area Relationships in Ecology," *Proceedings of the National Academy of Sciences* 103 (2006): 10310–15.

122 *wrote a paper with Phil*: Katherine D. McMahon, Héctor García Martín, and Philip Hugenholtz, "Integrating Ecology into Biotechnology," *Current Opinion in Biotechnology* 18 (2007): 287–92.

122 *metabolic flux analysis*: Amit Ghosh, Jerome Nilmeier, Daniel Weaver et al., "A Peptide-Based Method for ^{13}C Metabolic Flux Analysis in Microbial Communities," *PLoS Computational Biology* 10 (2104): e1003827.

123 *He gave me a preprint of an essay*: Nigel Goldenfeld and Carl Woese, "Life Is Physics: Evolution as a Collective Phenomenon Far from Equilibrium," *Annual Review of Condensed Matter Physics* 2 (2011): 375–99.

124 *I wrote op-eds*: Margonelli, "A Spill of Our Own," *New York Times*, May 1, 2010.

124 *article about the invention of nitrogen fertilizer*: Fields, "Global Nitrogen."

125 *mentioned in the Treaty of Versailles*: David R. Montgomery, *Dirt: The Erosion of Civilizations* (Berkeley: University of California Press, 2007; repr. [new preface] 2012), 196.

125 *allowing the world's population to grow*: Elizabeth Kolbert, "Head Count," *The New Yorker*, October 21, 2013, www.newyorker.com/magazine/2013/10/21/head-count-3.

125 *half the nitrogen in our bodies*: Sarah Zhang, "A Chemical Reaction Revolutionized Farming 100 Years Ago. Now It Needs to Go," *Wired*, May 16, 2016, www.wired.com/2016/05/chemical-reaction-revolutionized-farming-100-years-ago-now-needs-go/.

13. RESTLESS STREAMS

Muskrats were my dad's obsession more than mine, but their role as ecosystem engineers and their social nature make them part of this story. Not only have the muskrats worked with beavers to change the landscape of North America—adding a biological component to what we tend to see as geography—but their apparent cooperation also has a social aspect. Video by the researcher Cy Mott shows muskrats living in beaver dens in southern Illinois. Though they use different entrances, they seem—superficially, at least—to get along. See Cy L. Mott, Craig K. Bloomquist, and Clayton K. Nielsen, "Within-Lodge Interactions Between Two Ecosystem Engineers, Beavers (*Castor canadensis*) and Muskrats (*Ondatra zibethicus*)," *Behaviour* 150 (2013): 1325–44.

129 *remove some of his immune cells*: This was a study called "Vaccination of Patients with Renal Cell Cancer with Dendritic Cell Tumor Fusions and GM-CSF," with the clinical trials government identifier NCT00458536. It's listed at www.clinicaltrials.gov/ct2/show/NCT00458536.

129 *in a pattern resembling a clock*: The name of this drug was Granulocyte Macrophage Colony Stimulating Factor, and the injection site needed to be rotated, which was why we were told to inject in the pattern of a "clock."

129 *when they cut out a kidney tumor*: This is much less common than we understood at the time. Kidney tumors suppress the immune system in unknown ways. Very rarely—less than 1 percent of the time—when they are removed, lung tumors will spontaneously disappear. Suresh Bhat, "Role of Surgery in Advanced/Metastatic Renal Cell Carcinoma," *Indian Journal of Urology* 26 (2010): 167–76.

129 *His calcium levels*: Hypercalcemia is a symptom of some cancers, and it also has physical and psychological effects.

129 *"It took the form of a terrific onslaught"*: Eugène Marais, *Soul of the White Ant*, 138.

130 *special hemoglobin*: Robert A. MacArthur, Gillian L. Weseen, and Kevin L. Campbell, "Diving Experience and the Aerobic Dive Capacity of Muskrats: Does Training Produce a Better Diver?," *Journal of Experimental Biology* 206 (2003): 1153–61.

131 *Work ceased all over*: Eugène Marais, *Soul of the White Ant*, 153–54.

14. CROSSING THE ABSTRACTION BARRIER

I'd been following termites for a few years when a friend gave me a copy of *The Glass Bees*. Like Marais's scattershot analysis of termites, *The Glass Bees* was not an orderly dissection of how technology, possibility, and morality interact— it was nothing you'd find in a white paper, for example. But it still made a lot of intuitive sense. And I was surprised to find that the main character's experience lined up so perfectly with the experience of European military elites described by John Ellis in *The Social History of the Machine Gun*. Initially I resolved to ignore it in the greater project of a book about termites, but it kept coming back into my thoughts and giving me new ways to frame questions. Jünger was also a well-known entomologist who left behind a collection of forty thousand beetles.

135 *computer programming joke*: The use of this phrase as a "joke" can be found on Radhika's door and on various websites. The meaning of the phrase can be found in Harold Abelson and Gerald Jay Sussman with Julie Sussman, *Structure and Interpretation of Computer Programs*, 2nd ed. (Cambridge, Mass.: MIT Press, 1996), in the entry "Abstraction Barriers." See it online at https://mitpress.mit.edu/sicp/full-text/sicp/book/node29.html.

136 *the TERMES robots*: Here is a video of the project as it looked in 2011, from the Wyss Institute: www.youtube.com/watch?v=MmQOEaXIAnU.

136 *a challenge posed in 1995*: Guy Théraulaz and Eric Bonabeau, "Coordination in Distributed Building," *Science* 269 (1995): 686–88.

142 *a piece of the code*: Bill Gates, "Bill Gates Interview," interviewed by David Allison, Division of Computers, Information, and Society, National Museum of American History, Smithsonian Institution. Transcript available here: http://americanhistory.si.edu/comphist/gates.htm#tc11.

142 *a floating decimal to do math problems*: "The Math Package," from an annotated disassembly of the Altair BASIC 3.2 (4K). Found at http://altairbasic .org/math_ex.htm.

142 *they had the Kilobot*: Michael Rubenstein, Christian Ahler, and Radhika Nagpal, "Kilobot: A Low Cost Scalable Robot System for Collective Behaviors," *Proceedings of 2012 IEEE International Conference on Robotics and Automation (ICRA)* (2012): 3293–98.

143 *the same way that fireflies synchronize*: Steven H. Strogatz, *Sync: The Emerging Science of Spontaneous Order* (New York: Hyperion, 2003), 15–32.

144 *to build a self-assembling starfish*: Michael Rubenstein, Alejandro Cornejo, and Radhika Nagpal, "Programmable Self-Assembly in a Thousand-Robot Swarm," *Science* 345 (2014): 795–99.

145 *a science fiction book from the late 1950s*: Ernst Jünger, *The Glass Bees*, trans. Louise Bogan and Elizabeth Mayer, introduction by Bruce Sterling (New York: New York Review Books, 2000), 52.

146 *the same day I sat in traffic*: Agence France Presse, "US Drone Strike Kills Five Insurgents Near Afghan Border," *The Telegraph*, October 10, 2012, www .telegraph.co.uk/news/worldnews/asia/pakistan/9597737/US-drone -strike-kills-five-insurgents-near-Afghan-border.html.

15. INFLUENTIAL INDIVIDUALS

Another way to look at the problem of thinking without a brain is to think of it as a collective project. Iain Couzin and his lab have been studying the collective behaviors of locusts, fish, insects, and mammals both in the wild and in computer and mathematical models to understand how groups of animals without any special knowledge start to act smart. He has published papers with titles like "Consensus Decision-Making by Fish" and "The Social Context of Cannibalism in Migratory Bands of the Mormon Cricket," but for a more general sense of how he looks at schools of fish as systems of individuals that are both highly aware of each other and not so hyperaware that they get confused, read his short and elegant paper "Collective Minds," *Nature* 445 (2007): 715.

150 *more than two hundred hours*: Kirstin Hagelskjaer Petersen, "Collective Construction by Termite-Inspired Robots" (PhD dissertation, Harvard University, 2013), 90, http://nrs.harvard.edu/urn-3:HUL.InstRepos: 13068244.

151 *Anna Dornhaus's lab . . . tracked*: Daniel Charbonneau and Anna Dornhaus, "Workers 'Specialized' on Inactivity: Behavioral Consistency of Inactive Workers and Their Role in Task Allocation," *Behavioral Ecology and Sociobiology* 69 (2015): 1459–72.

151 *social insects must be industrious*: Michael Breed, "Why Are Workers Lazy?," *Insectes Sociaux* 62 (2015): 7–8.

152 *Iain Couzin has done extensive work*: Iain D. Couzin, "Collective Cognition in Animal Groups," *Trends in Cognitive Sciences* 13 (2009): 36–43.

154 *"useful not only as a matter of theoretical interest but . . ."*: Application for NIH grant, "Modeling Individual-to-Collective Behavior in Mound-Building Termites," project number 5R01GM112633-02, https://projectreporter.nih .gov/project_info_description.cfm?icde=0&aid=8927664.

154 *been recorded telling an audience*: Justin told his story for the project Story Collider. The recording can be found here: https://soundcloud.com/the -story-collider/justin-werfel-robotics-lessons/sets.

155 *studying how termites dig tunnels*: Paul Bardunias and Nan-Yao Su, "Queue Size Determines the Width of Tunnels in the Formosan Subterranean Termite (Isoptera: Rhinotermitidae)," *Journal of Insect Behavior* 23 (2010): 189–204.

155 *ancient Greek warfare*: Paul M. Bardunias and Fred Eugene Ray, Jr., *Hoplites at War: A Comprehensive Analysis of Heavy Infantry Combat in the Greek World, 750–100 B.C.E.* (Jefferson, N.C.: McFarland and Company, 2016).

155 *While Kirstin pulled insights out of the dataset*: Kirstin Petersen, Paul Bardunias, Nils Napp et al., "Arrestant Property of Recently Manipulated Soil on *Macrotermes michaelseni* as Determined Through Visual Tracking and Automatic Labeling of Individual Termite Behaviors," *Behavioural Processes* 116 (2015): 8–11.

156 *to understand that the cue for building*: Ben Green, Paul Bardunias, J. Scott Turner et al., "Excavation and Aggregation as Organizing Factors in De Novo Construction by Mound-Building Termites," *Proceedings of the Royal Society B: Biological Sciences* 284 (2017): pii 20162730.

16. THE ROBOT APOCALYPSE

So what actually happened when Reverend Billy came to Radhika's office? On the one hand, the protesters had a concept of who the scientists "are," and what their agenda is. Initially, the protesters inaccurately believed the bee research was supported by Monsanto, but later they amended that. Still, they did not treat the scientists as individuals or as people with whom they could talk in a normal way. Ironically, the protesters mirrored a recent process whereby scientists and science communicators have constructed ideas of who "the public" is. One reoccurring belief is that the public is superstitious and poorly informed and if allowed to have their way, this story goes, the public will block science. These mutual beliefs get in the way of starting a dialogue about such important topics. The sociologist Claire Marris has written extensively about this issue. Start with her essay "The Construction of Imaginaries of the Public as a Threat to Synthetic Biology," *Science as Culture* 24 (2015): 83–98.

160 *submitted a paper about their TERMES robots*: Justin Werfel, Kirstin Petersen, and Radhika Nagpal, "Designing Collective Behavior in a Termite-Inspired Robot Construction Team," *Science* 343 (2014): 754–58.

161 *"colonizing Mars"*: Doug Gross, "At Harvard, Swarming Robots That Mimic Termites," *CNN*, May 29, 2014, http://www.cnn.com/2014/05/29/tech/innovation/big-idea-swarm-robots/index.html.

161 *"Let's just hope those"*: "What Does the Future of Construction Look Like?," *Business Opportunities.biz*, January 19, 2015, http://www.business-opportunities.biz/2015/01/19/what-does-the-future-of-construction-look-like/.

161 *video on* Bloomberg News: Previously accessed at www.bloomberg.com/news/videos/b/e9579ba8-808c-49ad-b40d-e00714f1f14f (no longer available).

161 *Michael Crichton's 2002 bestseller* Prey: Michael Crichton, *Prey* (New York: HarperCollins, 2002).

162 *Radhika had painted*: You can see some of her paintings at her website: www.radhikanagpal.org/art.html.

164 *an essay in* Scientific American: Radhika Nagpal, "The Awesomest 7-Year Postdoc or: How I Learned to Stop Worrying and Love the Tenure-Track Faculty Life," *Scientific American Guest Blog,* July 21, 2013, https://blogs.scientificamerican.com/guest-blog/the-awesomest-7-year-postdoc-or-how-i-learned-to-stop-worrying-and-love-the-tenure-track-faculty-life/.

165 *I recognized Reverend Billy*: Jill Lane, "Reverend Billy: Preaching, Protest, and Postindustrial Flânerie," *TDR/The Drama Review* 46 (2002): 60–84.

166 *"funded by Monsanto"*: Elizabeth Kolbert, "Buzzkill," *The New Yorker*, May 19, 2014.

166 *Monsanto is a former chemical company*: Reuters, "Timeline of History of Monsanto," www.reuters.com/article/food-monsanto/corrected-timeline-history-of-monsanto-co-idUSN1032100920091111.

167 *I watched a video of the performance later*: www.youtube.com/watch?t=19&v=PYlaVO0tQPo.

17. DARWIN'S TERMITES

I found Australia's landscape and animals to be almost overwhelmingly weird. For the first few days I felt I couldn't see what I was looking at. Eventually, reading Bill Gammage's beautifully illustrated and extensively researched book *The Biggest Estate on Earth: How Aborigines Made Australia* (Sydney and London: Allen and Unwin, 2011) made the place more real to me.

174 Mastotermes darwiniensis, *the mastodon of termites*: Thomas Bourguignon, Nathan Lo, Stephen L. Cameron et al, "The Evolutionary History of Termites as Inferred from 66 Mitochondrial Genomes," *Molecular Biology and Evolution* 32 (2015): 406–21.

174 *"poster protist"*: Lynn Margulis and Dorion Sagan, *Dazzle Gradually: Reflections on the Nature of Nature* (New York: Sciencewriters/Chelsea Green, 2007), 45.

174 *a reputation as a natural cyborg*: See, for example, Donna Jeanne Haraway and Thyrza Nichols Goodeve, *How Like a Leaf: An Interview with Thyrza Nichols Goodeve* (New York: Routledge, 2000), 83.

175 *By 2070, more than three hundred days a year*: Australian Government, Department of the Environment and Energy, "Climate Change Impacts in the Northern Territory," www.environment.gov.au/climate-change/climate-science-data/climate-science/impacts/nt.

176 *found in amber*: See listings here: http://tolweb.org/tree?group=Mastotermitidae.

176 *even Wyoming*: Kenneth S. Bader, Stephen T. Hasiotis, and Larry D. Martin, "Application of Forensic Science Techniques to Trace Fossils on Dinosaur Bones from a Quarry in the Upper Jurassic Morrison Formation, Northeastern Wyoming," *Palaios* 24, no. 3 (2009): 140–58.

176 *speculated that climate change*: Grimaldi, *Evolution of the Insects*, 244.

177 *Great Barrier Reef, the world's largest*: Oliver Milman, "Climate Change Will Hit Australia Harder Than Rest of World, Study Shows," *The Guardian*, January 26, 2015.

177 *seventeen hundred plants and animals on the endangered*: www.australianwildlife.org/wildlife.aspx.

177 *"If the last blue whale choked"*: Quoted in O'Malley, *Philosophy of Microbiology*, 205.

177 *factories that produced antibiotics in India*: Erik Kristiansson, Jerker Fick, Anders Janzon et al., "Pyrosequencing of Antibiotic-Contaminated River Sediments Reveals High Levels of Resistance and Gene Transfer Elements," *PLoS ONE* 6 (2011): e17038.

179 *Recent genetic work suggests that they arrived:* Timothy R. C. Lee, Stephen L. Cameron, Theodore A. Evans et al., "The Origins and Radiation of Australian *Coptotermes* Termites: From Rainforest to Desert Dwellers," *Molecular Phylogenetics and Evolution* 82, part A (2015): 234–44.

180 *termites may use the Earth's magnetic field*: Peter Jacklyn and Ursula Munro, "Evidence for the Use of Magnetic Cues in Mound Construction by the Termite *Amitermes meridionalis* (Isoptera: Termitinae)," *Australian Journal of Zoology* 50 (2002): 357–68.

180 *expert Peter Jacklyn studied*: Peter Jacklyn, "Investigations into the Building Behaviour of a Minor Celebrity Insect," *Australian Zoologist* 35 (2010): 183–88.

180 Nasutitermes *mostly live in trees elsewhere*: Daej A. Arab, Anna Namyatova, Theodore A. Evans et al., "Parallel Evolution of Mound-Building and Grass-Feeding in Australian Nasute Termites," *Biology Letters* 13 (2017): 20160665.

181 *80 percent of the eucalyptus trees here*: Hizbullah Jamali, Stephen J. Livesley, Samantha P. Grover et al., "The Importance of Termites to the CH_4 Balance of a Tropical Savanna Woodland of Northern Australia," *Ecosystems* 14 (2011): 698–709.

181 *the trees burn differently:* Jason Beringer, Lindsay B. Hutley, David Abramson et al., "Fire in Australian Savannas: From Leaf to Landscape," *Global Change Biology* 21 (2015): 62–81.

182 *Captain Cook:* Gammage, *Biggest Estate on Earth,* 179.

182 *"a parallel universe" of protists:* Peter O'Donoghue, "Kingdom PROTISTA Haeckel, 1886," Australian Faunal Directory, Australian Biological Resources Study, November 2014, https://biodiversity.org.au/afd/taxa /PROTISTA?

183 *"poster animal":* Lynn Margulis and Dorion Sagan, "The Beast with Five Genomes," *Natural History Magazine,* June 2001.

183 *a way to use silver to make the protozoa visible:* Linda Ly, Coralyn Turner, Stephen Cameron, and Peter O'Donoghue, "Taxonomy of Endosymbiotic Protozoa of Native Termites from Northeastern Australia," in *Australian Society for Parasitology Annual Scientific Meeting,* Cairns, Australia, July 10–13, 2011.

183n *Margulis looked at the microbial world:* Andrew H. Knoll. "Lynn Margulis, 1938–2011," *Proceedings of the National Academy of Sciences* 109 (2012): 1022.

183n *"The speed, volume, and antiquity":* Lynn Margulis, *Acquiring Genomes: A Theory of the Origins of Species* (New York: Basic Books, 2002), 85.

184 *like longish ridged starfruit:* Brian S. Leander and Patrick J. Keeling, "Symbiotic Innovation in the Oxymonad *Streblomastix strix,*" *Journal of Eukaryotic Microbiology* 51 (2004): 291–300.

185 *"an increasingly strange mutational circus":* Patrick J. Keeling, John P. McCutcheon, and W. Ford Doolittle, "Symbiosis Becoming Permanent: Survival of the Luckiest," *Proceedings of the National Academy of Sciences* 112 (2015): 10101–103.

185 *"as though we are ourselves experimental subjects":* Patrick J. Keeling, "The Impact of History on Our Perception of Evolutionary Events: Endosymbiosis and the Origin of Eukaryotic Complexity," *Cold Spring Harbor Perspectives in Biology* 6 (2014): a016196.

18. THE SOUL OF THE SOIL

This is a good time to stop and talk about dirt. Dirt is amazing—a weird combination of inert stuff and a thing that is itself alive. And, of course, it appears to be one of the termites' main projects. For a very readable treatment of the subject, try *Dirt: The Erosion of Civilizations,* by David R. Montgomery (cited above, in the notes to chapter 12).

Also, if you have a sense that you have seen the story of the mine and the fight for its land, perhaps in a dream, it may be that you saw Werner Herzog's movie *Where the Green Ants Dream,* which was based loosely on the Gove Land Rights Case and included some members of the community in the cast. This essay, by Andrew W. Hurley, gets at some of the controversy and tension between the Yolngu community and the movie: "Whose Dreaming? Intercultural Appropriation, Representations of Aboriginality, and the Process of Film-Making in Werner Herzog's *Where the Green Ants Dream* (1983)," *Studies in Australasian Cinema* 1 (2007): 175–90.

186 *production company for* Mad Max: Fury Road *announced*: Noah Schultz-Byard, "'Mad Max: Fury Road' not filming in Broken Hill," *ABC*, August 4, 2011, www.abc.net.au/local/stories/2011/08/04/3285470.htm; Matt Buchanan and Scott Ellis, "Broken Hill Too Green for Mad Max," *Sydney Morning Herald*, August 5, 2011, http://www.smh.com.au /entertainment/movies/broken-hill-too-green-for-mad-max-20110804 -1idin.html.

187 *a paper about a remote bauxite mine*: Alister V. Spain, D. A. Hinz, J. A. Ludwig et al., "Mine Closure and Ecosystem Development—Alcan Gove Bauxite Mine, Northern Territory, Australia," in *Mine Closure 2006*, Andy Fourie and Mark Tibbett, eds., proceedings of First International Seminar on Mine Closure, Perth, Australia, September 13–15, 2006, 299–308.

187 *a farmer had increased the amount of wheat*: Theodore A. Evans, Tracy Z. Dawes, Philip R. Ward, and Nathan Lo, "Ants and Termites Increase Crop Yield in a Dry Climate," *Nature Communications* 2 (2011): 262.

187 *In 2050, as the population of the planet peaks*: Nikos Alexandratos and Jelle Bruinsma, "World Agriculture Towards 2030/2050: The 2012 Revision," ESA Working Paper No. 12–03 (Rome: FAO, 2012), 7.

187 *the amount of green stuff we grow per acre*: Hermann Lotze-Campen, Alexander Popp, Jan Philipp Dietrich, and Michael Krause, "Competition Between Food, Bioenergy and Conservation," Background Note to the World Development Report 2010 (Washington, D.C.: World Bank Group, 2012), http://siteresources.worldbank.org/INTWDR2010/Resources/5287678 -1255547194560/WDR2010_BG_Note_Lotze-Campen.pdf.

188 *fourteen thousand tons of sunscreen lotion*: C. A. Downs, Esti Kramarsky-Winter, Roee Segal et al., "Toxicopathological Effects of the Sunscreen UV Filter, Oxybenzone (Benzophenone-3), on Coral Planulae and Cultured Primary Cells and Its Environmental Contamination in Hawaii and the U.S. Virgin Islands," *Archives of Environmental Contamination and Toxicology* 70 (2016): 265–88.

189 *He had spent much of his career*: Alister is the coauthor of a classic book on soil ecology: Patrick Lavelle and Alister V. Spain, *Soil Ecology* (Dordrecht, The Netherlands: Kluwer Academic, 2001).

190 *forest destroyed by bauxite mining*: David Tongway and Tein McDonald, "Understanding the Less Visible Components for 'the Wise Management of Our Lands': Interview with David Tongway," *Ecological Management and Restoration* 17 (2016): 93–101.

190 *The bark petition*: Part of the collection of the Museum of Australian Democracy: www.foundingdocs.gov.au/item-sdid-100.html.

190 *"It seems to me that Yolngu perceive time as circular"*: Nancy M. Williams, *The Yolngu and Their Land: A System of Land Tenure and the Fight for Its Recognition* (Canberra: Australian Institute of Aboriginal Studies, 1986), 30.

191 *the land owned them*: John Hookey, "The Gove Land Rights Case: A Judicial

Dispensation for the Taking of Aboriginal Lands in Australia?," *Federal Law Review* 5 (1972): 85–114; 103.

193 *twelve kinds of termites*: Alister V. Spain, Mark Tibbett, Dieter A. Hinz et al., "The Mining-Restoration System and Ecosystem Development Following Bauxite Mining in a Biodiverse Environment of the Seasonally Dry Tropics of Australia," in *Mining in Ecologically Sensitive Landscapes*, ed. Mark Tibbett (Melbourne: CSIRO Publishing, 2015), 173.

19. THE MATH OF FAIRY CIRCLES

Before I knew about Turing patterns or patterned ecosystems, I read a paper that made the rounds of people concerned about climate change and ecological and social stability in 2009, called "Early-Warning Signals for Critical Transitions," by Marten Scheffer, Jordi Bascompte, William A. Brock et al., *Nature* 461: 53–59. The paper proposed that ecosystems and financial systems go through "critical transitions," where they cease being reasonably resilient and, instead, crash. Perhaps because we were all dealing with the aftermath of 2008's financial crash, which in hindsight looked predictable, the idea that there might be predictable changes that could be observed across natural, social, and financial systems was very attractive. As this book went to press, the paper had been cited nearly 1,900 times.

When I read Corina and Rob's paper on termites providing resilience in dry ecosystems, I remembered the critical transitions paper and realized that in hindsight I really hadn't understood the theories it was based on. These cross-disciplinary studies—ecology, biological mathematics, complexity science, and other fields—only make it out of academia occasionally. I suspect this one did because it suggested predictability in a time when the future felt precarious.

196 *2010, when he and a team published a paper*: Robert M. Pringle, Daniel F. Doak, Alison K. Brody et al., "Spatial Pattern Enhances Ecosystem Functioning in an African Savanna," *PLoS Biology* 8 (2010): e1000377.

196 *termites had organized the entire landscape from below*: Pascal Jouquet, Saran Traoré, Chutinan Choosai et al., "Influence of Termites on Ecosystem Functioning. Ecosystem Services Provided by Termites," *European Journal of Soil Biology* 47 (2011): 215–22.

197 *termites' influence had to do with nutrients*: Kena Fox-Dobbs, Daniel F. Doak, Alison K. Brody, and Todd M. Palmer, "Termites Create Spatial Structure and Govern Ecosystem Function by Affecting N_2 Fixation in an East African Savanna," *Ecology* 91 (2010): 1296–1307.

197 *a series of feedback loops*: P. Jens Dauber, Jan Lagerlöf et al., "Soil Invertebrates as Ecosystem Engineers: Intended and Accidental Effects on Soil and Feedback Loops," *Applied Soil Ecology* 32 (2006): 153–64.

197 *LIDAR (Light Detection and Ranging)*: Shaun R. Levick, Gregory P. Asner,

Oliver A. Chadwick et al., "Regional Insight into Savanna Hydrogeomorphology from Termite Mounds," *Nature Communications* 1 (2010): 65.

198 *on a farm in Romania*: Karen A. Frenkel, "Women in Science: Their Personal Journeys," *News from Science*, June 3, 2011, www.sciencemag.org/news/2011/06/women-science-their-personal-journeys.

198 *high-dimensional geometry*: Jessica P. Johnson, "Corina Tarnita: The Ant Mathematician," *The Scientist* 25 (2011): 51.

198 *In 2010 she published the big paper*: Martin A. Nowak, Corina E. Tarnita, and Edward O. Wilson, "The Evolution of Eusociality," *Nature* 466 (2010): 1057–62.

199 *resembled a Turing pattern*: Max Rietkerk and Johan van de Koppel, "Regular Pattern Formation in Real Ecosystems," *Trends in Ecology and Evolution* 23 (2008): 169–75.

199 *the way rabbit brains perceive smells*: Bill Baird, "Nonlinear Dynamics of Pattern Formation and Pattern Recognition in the Rabbit Olfactory Bulb," *Physica D: Nonlinear Phenomena* 22 (1986): 150–75.

200 *a patchwork of hexagons*: Robert M. Pringle and Corina E. Tarnita, "Spatial Self-Organization of Ecosystems: Integrating Multiple Mechanisms of Regular-Pattern Formation," *Annual Review of Entomology* 62 (2017): 359–77; 367.

202 *termite mounds made the landscape much more drought resistant*: Juan A. Bonachela, Robert M. Pringle, Efrat Sheffer et al., "Termite Mounds Can Increase the Robustness of Dryland Ecosystems to Climatic Change," *Science* 347 (2015): 651–55.

202 *models, from the mid-2000s*: Max Rietkerk, Stefan C. Dekker, Peter C. de Ruiter, and Johan van de Koppel, "Self-Organized Patchiness and Catastrophic Shifts in Ecosystems," *Science* 305 (2004): 1926–29; Sonia Kéfi, Max Rietkerk, Concepción L. Alados et al., "Spatial Vegetation Patterns and Imminent Desertification in Mediterranean Arid Ecosystems," *Nature* 449 (2007): 213–17.

204 *A small spate of papers attributing the circles*: Cristian Fernández-Oto, Mustapha Tlidi, D. Escaff, and Marcel Gabriel Clerc, "Strong Interaction Between Plants Induces Circular Barren Patches: Fairy Circles," *Philosophical Transactions of the Royal Society A: Mathematical, Physical and Engineering Sciences* 372 (2014): 20140009; Stephan Getzin, Kerstin Wiegand, Thorsten Wiegand et al., "Adopting a Spatially Explicit Perspective to Study the Mysterious Fairy Circles of Namibia," *Ecography* 38 (2015): 1–11; Stephan Getzin, Hezi Yizhaq, Bronwyn Bell et al., "Discovery of Fairy Circles in Australia Supports Self-Organization Theory," *Proceedings of the National Academy of Sciences* 113 (2016): 3551–56.

205 *In 2017 the team . . . published a paper*: Corina E. Tarnita, Juan A. Bonachela, Efrat Sheffer et al., "A Theoretical Foundation for Multi-Scale Regular Vegetation Patterns," *Nature* 541 (2017): 398–401.

206 *the problem of scale in ecology*: Simon A. Levin, "The Problem of Pattern and Scale in Ecology: The Robert H. MacArthur Award Lecture," *Ecology* 73 (1992): 1943–67.

20. THE SOUL OF THE CELL

For a basic background on synthetic biology, I found Adam Rutherford's *Creation: How Science Is Reinventing Life Itself* (New York: Penguin, 2013) interesting and revealing of the way that the field talked about itself and its goals, particularly during the period I was in the labs. To get a deeper sense of why synthetic biologists talk about their work the way they do, the cultural anthropologist Sophia Roosth spent eight years talking with them. She started with those who were clustered around MIT but also went to UC Berkeley's Joint Bio-Energy Institute and beyond. She views the field slantwise, wondering how a field that felt it was making the world—and biology—anew came to use so many older ideas to talk about itself. *Synthetic: How Life Got Made* (Chicago: University of Chicago Press, 2017).

210 *By 2015, they had metabolic maps*: This can be found in a database called "Arrowland" at JBEI: https://public-arrowland.jbei.org/. For more information about the background of the project, see: Héctor García Martín, Vinay Satish Kumar, Daniel Weaver et al., "A Method to Constrain Genome-Scale Models with ^{13}C Labeling Data," *PLoS Computational Biology* 11 (2015): e1004363.

210 *nearly nineteen hundred peer-reviewed papers*: Steven C. Slater, Blake A. Simmons, Tamara S. Rogers et al., "The DOE Bioenergy Research Centers: History, Operations, and Scientific Output," *BioEnergy Research* 8 (2015): 881–96.

211 *the price of oil went nuts*: U.S. Energy Information Administration, Short-Term Energy Outlook, Real Prices Viewer: https://www.eia.gov/outlooks/steo/realprices/.

212 *one from Sydney Brenner*: Elizabeth Dzeng, "How Academia and Publishing Are Destroying Scientific Innovation: A Conversation with Sydney Brenner," *Kings Review*, February 24, 2014, http://kingsreview.co.uk/articles/how-academia-and-publishing-are-destroying-scientific-innovation-a-conversation-with-sydney-brenner/.

214 *one of the exotics was called pinene*: Stephen G. Sarria, Betty Wong, Héctor García Martín et al., "Microbial Synthesis of Pinene," *ACS Synthetic Biology* 3 (2014): 466–75.

214 *JP-10, an advanced rocket fuel*: Jim Lane, "9 Advanced Molecules That Could Revolutionize Jet and Missile Fuel," *Biofuels Digest*, June 18, 2014, http://www.biofuelsdigest.com/bdigest/2014/06/18/9-advanced-molecules-that-could-revolutionize-jet-and-missile-fuel/.

214 *the Defense Department's research arm DARPA ramped up*: Todd Kuiken, "U.S.

Trends in Synthetic Biology Research Funding," Wilson Center Synthetic Biology Project, September 15, 2015, www.wilsoncenter.org/publication /us-trends-synthetic-biology-research-funding.

214 *price of wormwood fell*: Mark Peplow, "Synthetic Biology's First Malaria Drug Meets Market Resistance," *Nature* 530 (2016): 389–90.

214 *visions of utopia and those of apocalypse*: Claire Marris and Nikolas Rose, "Let's Get Real on Synthetic Biology," *New Scientist* 214 (June 6, 2012): 28–29.

214 *Genomatica's 1,4 butanediol*: Press release: "BASF and Genomatica Expand License Agreement for 1,4-Butanediol (BDO) from Renewable Feedstock," September 24, 2015, www.basf.com/en/company/news-and-media/news -releases/2015/09/p-15-347.html.

215 *DuPont's 1,3 propanediol*: MarketsandMarkets, "1,3-Propanediol (PDO) Market by Applications (PTT, Polyurethane, Cosmetic, Personal Care & Home Cleaning & Others) & Geography-Global Market Trends & Forecasts to 2021," www.marketsandmarkets.com/Market-Reports/1-3-propanediol -pdo-market-760.html. Also: European Commission, "Environmental Factsheet: 1,3 Propanediol," https://ec.europa.eu/jrc/sites/jrcsh/files/BISO -EnvSust-Bioproducts-13PDO_140930.pdf.

215 *"An open question is whether biology"*: Jeffrey C. Way, James J. Collins, Jay D. Keasling, and Pamela A. Silver, "Integrating Biological Redesign: Where Synthetic Biology Came From and Where It Needs to Go," *Cell* 157 (2014): 151–61.

216 *disrupt the cell's internal communications*: Jens Nielsen and Jay D. Keasling, "Engineering Cellular Metabolism," *Cell* 164 (2016): 1185–97.

216 *By 2016, the team's work increased*: Personal communication with Héctor.

217 *create four-hundred-fold changes in production*: Victor Chubukov, Aindrila Mukhopadhyay, Christopher J. Petzold et al., "Synthetic and Systems Biology for Microbial Production of Commodity Chemicals," *NPJ Systems Biology and Applications* 2 (2016): 16009.

217 *"Whether biological systems can be understood"*: Ibid.

21. EMPATHY AND THE DRONE

As I started thinking more about how decisions about technology get made, and what democracy's role is, I read the Harvard ethicist Sheila Jasanoff's book *The Ethics of Invention: Technology and the Human Future* (New York: W. W. Norton, 2016). Jasanoff sees technology itself as a form of governance which we're often coerced into accepting. Her work is a good introduction to the many ethicists and thinkers who've worked on more-inclusive ways to set technological agendas and policy. For a thought-provoking analysis of the philosophical implications of drones and how they change the ethics of war, read Grégoire Chamoyou's *Theory of the Drone*, translated by Janet Lloyd (New York: New Press, 2014).

218 *as many as three Predator drones*: Personal communication from Thomas Barry, Senior Analyst, Center for International Policy, January 12, 2015,

confirming presence of drones in 2008; Thomas Barry, "Drones Over Homeland: Expansion of Scope and Lag in Governance," *Brown Journal of World Affairs* 19 (2013): 65–80.

218 *pilots in air-conditioned trailers back at Fort Huachuca*: Fort Huachuca is in Sierra Vista, Arizona. Pia Bergvist, "Drone Pilot Jobs: Flying a UAV," *Flying Magazine*, June 16, 2014, www.flyingmag.com/aircraft/drone-jobs-what -it-takes-fly-uav#page-5.

218 *from a height of twenty-two thousand feet they can "see"*: Brian Bennett, "Predator Drones Do Domestic Duty," *Los Angeles Times*, September 12, 2011, http://articles.latimes.com/2011/sep/12/nation/la-na-domestic-drones -20110912.

219 *By 2014, ten of them were flying*: Office of the Inspector General, Homeland Security, "U.S. Customs and Border Protection's Unmanned Aircraft System Program Does Not Achieve Intended Results or Recognize All Costs of Operations," December 24, 2014, 21, www.oig.dhs.gov/assets /Mgmt/2015/OIG_15-17_Dec14.pdf.

219 *engineer live insects as "allies"*: Blake Bextine, "Insect Allies," DARPA program information, www.darpa.mil/program/insect-allies.

219n *"It does not seem as though any influence"*: Sigmund Freud, *Civilization and Its Discontents*, trans. James Strachey (New York: W. W. Norton, 1962), 22.

220 *"As flying robots become smaller"*: Stuart Russell, "Take a Stand on AI Weapons," *Nature* 521 (2015): 416.

220 *"smart clouds"*: Paul Scharre, "Robotics on the Battlefield Part II: The Coming Swarm," Center for a New American Security, October 2014: 16, 22, 34, 27.

220 *In late 2014 the U.S. Navy*: Stew Magnuson, "Autonomous Machines to Defeat Threats Beyond the Speed of Thought," *National Defense*, November 1, 2014, www.nationaldefensemagazine.org/articles/2014/11/1/2014novem ber-autonomous-machines-to-defeat-threats-beyond-the-speed-of-thought.

220 *By 2016, a cloud*: "Department of Defense Announces Successful Micro-Drone Demonstration," Department of Defense press release NR-008-17, January 9, 2017, www.defense.gov/News/News-Releases/News-Release View/Article/1044811/department-of-defense-announces-successful-micro -drone-demonstration/.

221n *Fritz Haber converted* Gilbert King, "Fritz Haber's Experiments in Life and Death," *Smithsonian Magazine*, June 6, 2012, www.smithsonianmag.com /history/fritz-habers-experiments-in-life-and-death-114161301/?no-ist.; John Cornwell, *Hitler's Scientists: Science, War, and the Devil's Pact* (New York: Viking, 2003), chapter 4; Sarah Hazell, "Mustard Gas—from Great War to Frontline Chemotherapy," *Scienceblog* of Cancer Research UK, August 27, 2014, http://scienceblog.cancerresearchuk.org/2014/08/27 /mustard-gas-from-the-great-war-to-frontline-chemotherapy/.

223 *Laws of War* and *rules of engagement*: These crop up in slightly different versions in U.S. and international guidebooks. Before joining a conflict, U.S.

forces are often given a card that defines the rules of engagement for that theater. The entire set of rules is laid out in: Office of the General Counsel, Department of Defense, *Department of Defense Law of War Manual*, June 12, 2015, 51–52, http://archive.defense.gov/pubs/law-of-war-manual -june-2015.pdf.

224 *In 2012 the Department of Defense decided:* The DOD issued a ten-year directive: Directive 3000.09. See the directive at: https://www.law.upenn.edu /live/files/3404-dod-autonomy-in-weapon-systems-2012. For more information see Bonnie Lynn Docherty, "Losing Humanity: The Case Against Killer Robots," Human Rights Watch and the International Human Rights Clinic, Human Rights Program, Harvard Law School, November 2012.

224 *Some theorists have proposed that autonomous robot armies:* Robert C. Arkin, "Ethical Robots in Warfare," *IEEE Technology and Society Magazine* 28 (2009): 30–33.

225 *In 1899 more than two dozen countries met in The Hague:* Database, "Treaties, States Parties and Commentaries," International Committee of the Red Cross: https://ihl-databases.icrc.org/applic/ihl/ihl.nsf/Treaty .xsp?documentId=CD0F6C83F96FB459C12563CD002D66A1&action =openDocument.

226 *treaties have banned blinding lasers:* Human Rights Watch Memorandum to Convention on Conventional Weapons Delegates, "Precedent for Preemption: The Ban on Blinding Lasers as a Model for a Killer Robots Prohibition," November 8, 2015, https://www.hrw.org/news/2015/11/08/preced ent-preemption-ban-blinding-lasers-model-killer-robots-prohibition.

227 *such worrisome products as synthetic opioids:* Kenneth A. Oye, J. Chappell, H. Lawson, and Tania Bubela, "Drugs: Regulate 'Home-Brew' Opiates," *Nature* 521 (2015): 281–83.

227 *Less than 1 percent of the federal money:* Kuiken, Wilson Center, "U.S. Trends in Synthetic Biology Funding."

228 *"Trust in science is just as fragile":* Sheila Jasanoff, J. Benjamin Hurlbut, and Krishanu Saha, "CRISPR Democracy: Gene Editing and the Need for Inclusive Deliberation," *Issues in Science and Technology* 32 (2015): 37–49.

22. WHITE ANTS

The Yirrkala community has done an enormous amount of work to be assured that they can tell their own story. The movie *This Is My Thinking* is both an argument and an initiation: once you've seen it, the terms of the arrangement between the community and the mining company are forever changed.

Today the Buku-Larrnggay Mulka Art Centre maintains an active archive of video and audio records from the community. You can see them at https:// yirrkala.com/video-the-mulka-project/.

230 *Six months before:* James Wilson, "Rio Tinto to Close Australian Alumina Refinery," *Financial Times*, November 29, 2013.

232 *a deliberate strategy*: Philippa Deveson, "The Agency of the Subject: Yolngu Involvement in the Yirrkala Film Project," *Journal of Australian Studies* 35 (2011): 153–64.

233 *"lifestyle choice" of living in remote towns*: Emma Griffiths, "Indigenous Advisers Slam Tony Abbott's 'Lifestyle Choice' Comments as 'Hopeless, Disrespectful,'" *ABC*, March 11, 2015, www.abc.net.au/news/2015-03-11/abbott -defends-indigenous-communities-lifestyle-choice/6300218.

235 *created by Wukun Wanambi*: National Museum Australia, "Unsettled" exhibition contributor note on Wukun Wanambi: www.nma.gov.au /exhibitions/unsettled/wukun_wanambi.

235 *"at this time"*: Email from Rio Tinto PR person to the author, July 14, 2014.

236 This Is My Thinking: Ian Dunlop, Chris Oliver, Philippa Deveson, Film Australia, Australian Institute of Aboriginal and Torres Strait Islander Studies et al., *This Is My Thinking*, ©2011 National Film and Sound Archive of Australia. You can order the film from http://shop.nfsa.gov.au /this-is-my-thinking.

236 *Another Dunlop film*: Ian Dunlop, Chris Oliver, Philippa Deveson, Film Australia, Australian Institute of Aboriginal and Torres Strait Islander Studies et al., *Pain for This Land: Footage from the Early Years, 1970–1971*, © 2011 National Film and Sound Archive/Film Australia Collection, http:// shop.nfsa.gov.au/pain-for-this-land~4926.

237 *son of Milirrpum Marika*: Here is a sketch of the family as artists from the Australian National Museum: www.nma.gov.au/exhibitions/yalangbara /the_marika_family.

238 *I had seen Wanyubi talking*: "Yirrkala Drawings: Wanyubi Marika," Art Gallery New South Wales, video: https://youtu.be/kcOcDPMqSAg.

239 *the Garma Festival gave them social influence*: Pauline Moullot, "Whither the Aborigine?," *World Policy Journal*, Fall 2015, www.worldpolicy.org/journal /fall2015/whither-aborigine.

239 *"Constructing a 'Futurology from Below'"*: Jim Thomas, "Constructing a 'Futurology from Below': A Civil Society Contribution Toward a Research Agenda," *Journal of Responsible Innovation* 2 (2015): 92–95.

241 *72 percent of the royalties*: Amos Aikman, "Rirratjingu Clan Ready to Sign Historic Lease Over Yirrkala," *The Australian*, September 15, 2014.

241 *knocked over an even more famous banyan tree*: Melanie Wilkinson, Dr. R. Marika, and Nancy M. Williams, "This Place Already Has a Name," in *Aboriginal Placenames: Naming and Re-naming the Australian Landscape*, ed. Harold Koch and Luise Hercus, Aboriginal History Monograph Series 19 (Canberra: ANU E Press and Aboriginal History Inc., 2009), 403–62.

241 *sent another bark petition to Parliament*: The petition, submitted in 1968, is described in this Australian Government article titled "Bark Petitions: Indigenous Art and Reform for the Rights of Indigenous Australians": www.australia.gov.au/about-australia/australian-story/bark-petitions -indigenous-art.

23. *THEM* AND US

I spent many hours following the physicists Hunter King and Sam Ocko around the fields as they inserted delicate handmade instruments into the mounds to measure the way gas and heat flowed during that 2014 trip. Frequently, shortly after they had gotten the instruments situated, they'd realize a termite was trying to "repair" the mounds with mud. By doing similar experiments in India, their work ultimately led to the realization that different termites, building different mounds, are creating functional structures that use the sun's energy in different ways to circulate heat and gases within the mound. (See citation below.) The Namibian mounds, for example, kept fairly steady concentrations of carbon dioxide inside them, while the Indian mounds, built by *Odontotermes obesus*, tended to have concentrations of the gas that rose higher during the day and then whooshed out after sunset. They concluded that the mounds do work like lungs and could offer interesting models for ventilating human-built structures.

248 *two physicists who were measuring gas and temperature flows*: Samuel A. Ocko, Hunter King, David Andreen et al., "Solar-Powered Ventilation of African Termite Mounds," *Journal of Experimental Biology* 220 (2017): 3260–69.

250 *a golden retriever has 627 million neurons*: Débora Jardim-Messeder, Kelly Lambert, Stephen Noctor, Fernanda M. Pestana, Maria E. de Castro Leal, et al., "Dogs Have the Most Neurons, Though Not the Largest Brain: Trade-Off Between Body Mass and Number of Neurons in the Cerebral Cortex of Large Carnivoran Species," *Frontiers in Neuroanatomy* 11 (2017): 5.

250 *a termite might have about*: Sarah M. Farris and Nicholas J. Strausfeld, "A Unique Mushroom Body Substructure Common to Basal Cockroaches and to Termites," *The Journal of Comparative Neurology*, 436 (2003): 306.

252 *Maeterlinck's* The Life of the White Ant: Maurice Maeterlinck, *The Life of the White Ant* (New York: Dodd, Mead, 1927).

254 *the two species have begun to hybridize*: Thomas Y. Chouvenc, Ericka E. Helmick, and Nan-Yao Su, "Hybridization of Two Major Termite Invaders as a Consequence of Human Activity," *PLoS ONE* 10 (2015): e0120745.

254 *most likely to succeed*: Grzegorz Buczkowski and Cleo Bertelsmeier, "Invasive Termites in a Changing Climate: A Global Perspective," *Ecology and Evolution* 7 (2017): 974–85.

255 *when woodlands in Sumatra*: David T. Jones, F. X. Susilo, David E. Bignell et al., "Termite Assemblage Collapse Along a Land-Use Intensification Gradient in Lowland Central Sumatra, Indonesia," *Journal of Applied Ecology* 40 (2003): 380–91.

255 *A group of naturalists in Germany*: Gretchen Vogel, "Where Have All the Insects Gone?," *Science* 356 (2017): 576–79.

255 *one global index measured a 45 percent decline*: Rodolfo Dirzo, Hillary S. Young, Mauro Galetti et al., "Defaunation in the Anthropocene," *Science* 345 (2014): 401–406.

255 *only 1 million of the 5 million*: Robert R. Dunn, "Modern Insect Extinctions, the Neglected Majority," *Conservation Biology* 19 (2005): 1030–36.

288

ACKNOWLEDGMENTS

Letting a reporter come into your lab for a few hours is very different from letting one come into your life for days, weeks, and years. The principal investigators in this book were beyond generous: Scott Turner, Phil Hugenholtz, Radhika Nagpal, Héctor Garcia Martin, Rob Pringle, and Corina Tarnita. If not for their deep willingness to share their knowledge and process—as well as their frustrations—this book would not exist.

Once I was in the labs, I hugely appreciated the time and attention of many researchers who tolerated me hanging around while they worked, often going to great lengths to accommodate my presence as well as my ignorance: Anna Engelbrektson, Shaomei He, Nurdyana Abdul Rahman, Rochelle Soo, Nancy Lachner, Ben Robedee, Vincent Mughonghora, Justin Werfel, Nils Napp, Kirstin Petersen, Paul Bardunias, Hunter Evens, Sam Ocko, Garrett Birkel, and many others from JBEI. The very first researcher I prevailed upon in this way was Falk Werneke, who passed away much too soon. Without their extraordinary patience, I wouldn't have understood a thing.

In Australia, Alister Spain and Dieter Hinz graciously told me about their work. In Arnhem Land, I was helped by many people, including Yalmay Yunupingu, Ritjilili Ganambarr, Banbapuy Ganambarr-Whitehead, Valerie Milminyina Dhamarrandji and her family, the artists and staff of the Buku-Larrngay Mulka Art Center, Will Stubbs and his family, Stuart Maclean of the Rirratjingu Aboriginal Corporation, Glenn Aitchison of Yolngu Business Enterprises, and Klaus Helms of the Gumatj Aboriginal Corporation.

Over the years, many people offered help that was both practical and insightful. Among those people were David Gilbert at JGI, Eugene Marais, Berry Pinchon, Rupert Soar, the staff of the Omatjenne Research Station, the staff at the Cheetah Conservation Fund, Rudi Schefferahn and friends, Brian Thistleton and Michael Neal of the Northern Territory Department of Primary Industry and Fisheries, Kevin Costa of Synberc, the International Union for the Study of Social Insects, Paul Eggleton, David Bignell, Jake Kosek, Megan Palmer, Dr. Daman S. Walia and the staff of ARCTECH, Dr. Don Ewart, Lindsay Hutley, Hizbullah Jamali, Jonathan Eisen, Mary Maxon, Manfred Auer, Robert Rapier, Peter Matlock, and Marcy Darnovsky.

Opportunities for research were provided by Arizona State University Library, Patten Free Library and State of Maine Interlibrary Loan, New York Public Library Manhattan Research Library Initiative, and the UC Berkeley Library.

People who encouraged termite performances, rants, and articles over the years: Doug McGray and Pop-Up Magazine, the New Year's Eve Party of Beth Lisick and Eli Crews, Mary Roach, Heather Smith, and Novella Carpenter. Don Peck, Maria Streshinsky, Barbara Paulsen, and Kerry Tremain assigned stories on termites and related subjects.

Dan Sarewitz and Taryn Brant Bowe did thoughtful reads of an early draft. Marc Herman, Nina Sazevich, Connie Voisine, and Becky Margonelli Haliscelik also read and offered many helpful suggestions.

Editors Amanda Moon and Sean McDonald encouraged me to write about more than "just" termites, and then helped sort out the aftermath. Essential editorial insights and all sorts of invaluable help were provided by Scott Borchert, Dominique Lear, and Daniel Vazquez. Copy editor Annie Gottlieb and production editor Susan Goldfarb then attended to every word and detail. My agent, Jay Mandel, believed in termites from the very beginning.

Gregory Rodriguez and my colleagues at Zócalo Public Square for giving me an intellectual home that termites can't eat.

Clarke Cooper for every semicolon.

INDEX

Abbott, Tony, 230

Aboriginals, Australian, 38, 59, 76–77, 182, 190, 221; *see also* Yolngu people

Aesop's Fables, 151

Afghanistan, 219, 222, 225

Africa, 7, 194, 199, 214, 236; European colonization of, 98, 265*n*; termites in, 5, 179, 196, 254; *see also* Namibia

Agriculture, U.S. Department of (USDA), 98

alates, 17

algorithms, 63, 137–42, 144–45, 161, 163, 181

Allen, Paul, 142

Allgaier, Martin, 107, 108, 112–13

Altair 8080, 142

altruism, 35–36, 39, 58, 110–11, 163, 206, 253

American Association for the Advancement of Science, 160

Amitermes, 110*n*, 193

ammonia, 111, 177

Amritsar (India), 72, 163

Angola, 15, 32, 59

Annual Review of Condensed Matter Physics, 123

"Ant-Colony as an Organism, The" (Wheeler), 22

Anthropocene epoch, 10

ants, 7–10, 36, 70, 151–52, 181, 198, 249; books about, 18; crazy, 236–37; green, 179; human admiration for, 9; leaf-cutter, 24; soil improved by, 187, 194; structural differences of termites and, 8; superorganisms of, 22; *see also* social insects

Apartheid, 29

apocalypse, 163–64, 214, 221, 228, 255; robot, 161, 224

Apple, 209

Arafura Sea, 179, 229

Arizona, 3, 5–12, 84, 108, 242; University of, 10, 151

Army, U.S., 57, 219

Arnhem Land (Australia), 230, 239–40

artemisinin, 104–105, 214

Asia, 5, 24, 179, 205, 254; *see also* *specific nations*

Atlantic, The, 4

atom bomb, 11, 24, 125

Australia, 5, 12, 115, 116, 173–95, 202, 229–42, 254, 275n; Aboriginal peoples of, 38, 59, 76–77, 182, 190, 221 (*see also* Yolngu people); bauxite mining in, 187, 190–94, 220, 229–31, 233, 235–40; climate change in, 186–87; fairy circles in, 204; protozoa in, 182–85; termite safari in, 174–81; in World War II, 175; *see also specific states, municipalities, and regions*

australopithecines, 17

automata, 23, 68–69, 146, 164, 254

bacteria, 97, 101, 106, 124, 192, 215; antibiotic-resistant genes evolving in, 177–78; gut, 40, 85–88, 108–11, 111; symbiosis of protozoa and, 183–85; see also *E. coli*

Banfield, Jill, 89

Bardunias, Paul, 155–57, 248–50

Bascompte, Jordi, 279n

BASIC computer language, 142

bauxite, 187, 189–91, 194, 229

beavers, 47, 156, 271n

Beckett, Samuel, 74, 253

bees, 8–10, 12, 18, 44, 183; robotic, 146 (*see also* RoboBees); *see also* social insects

benchwork, 97–98, 210

Bennett, Gaymon, 106

Berkeley (California), 127; *see also* California, University of, Berkeley

Bignell, David, 10

Bill and Melinda Gates Foundation, 104–105

biochemistry, 45, 86, 210

biofuels, 85–86, 90, 97–100, 104–105, 115–19, 122–26, 207–17, 219; *see also* grassoline

Bioinformatics for Dummies (Claverie), 95

biological weapons, 226

biology, 11, 15, 51, 54, 65, 86, 149, 253; of cognition, 57, 69; complex systems in, 49, 74, 154–55; of environmental clues, 47–48; evolutionary, 25, 35–36, 42–43; of fairy circles, 201, 202; fieldwork in, 78, 94; mathematical, 12, 28–29, 36, 119, 198–99, 253;

molecular, 99, 123; of nitrogen fixing, 108; physics and, 116–24, 216–17; physiological, 12, 29, 30, 51, 65; robotics and, 61–62, 71–72, 162–63, 167–68; of superorganisms, 23–24; of symbiogenesis, 185; synthetic, *see* synthetic biology; theoretical, 199–200; *see also* microbiology

bioreactors, 83, 85–86, 122, 217

black boxes, 60–72, 75, 86, 92–93, 121, 197, 204, 217

Black Dahlia, The (Ellroy), 27

BLASTs, 109

Bloomberg News, 161

blue-green algae, 183

bombs, 78, 118, 146, 175, 226; atom, 11, 24, 125

Bonabeau, Eric, 48, 135, 263n

Bonachela, Juan, 200–203

Botswana, 29

BP oil and gas company, 85, 104

Brave New World (Huxley), 23

Brazil, 6, 175, 205

Brenner, Sydney, 212

Brisbane (Australia), 174, 176, 177, 179, 182, 229

British Columbia, University of, 114

British Empire, 76

Brock, William A., 279n

Buku-Larrnggay Mulka Art Center (Yirrkala), 229, 284n

Bunuwal Group, 237

Burj Khalifa (Dubai), 31

butanediol (BDO), 214–15

calcium, 42, 92, 129

calcrete, 42

California, 5, 10, 12, 17, 130, 207–17; University of, Berkeley, 85, 89, 91n, 102, 103, 176, 220, 281n

Cape Town (South Africa), University of, 29

Carnot, Nicolas Léonard, 86, 158, 161, 196, 216, 253; Waiting for, 74, 102, 119

Cas9, 103–104

Castillo-Vardaro, Jessica, 200–203

Cell, 215

cellulose, 52, 87, 92, 97, 99–100; breaking down, 85, 108, 109, 123, 168, 208, 210

cement pheromones, 48–50, 78, 155–56

Central America, 5, 179

Chamoyou, Grégoire, 282n

Cheetah Conservation Fund, 260n

chemical weapons, 226; *see also* poison gases

chemistry, 102, 120–22, 125, 205, 216, 221n, 248, 251; *see also* biochemistry

Chicago, 160

China, 93, 214

chloroplasts, 183

Christianity, 221n

Clarke, Arthur C., 155, 159, 222

"Cleansing Patrols," 76, 221

climate change, 5, 163, 174–77, 181, 186–87, 203, 238, 254, 279n; oil and, 95, 125–26

Clostridium difficile, 116

clustered regularly interspaced palindromic repeats (CRISPRs), 103–104

Clusters of Orthologous Groups (COGs), 108

cockroaches, 6, 8, 174–76, 254
coevolution, 91, 115, 179
Cold War, 35
Colorado River, 11
Commonwealth Scientific and
    Industrial Research Organization
    (CSIRO), 181, 187, 189
computer science, 12, 49, 61, 63,
    71–72, 264n
computer simulations, 16, 54, 138,
    156
condensed matter physics, 119, 123,
    209
Congo Red dye, 98
Cook, Garry, 181
Cook, Captain James, 182, 188
Coomassie blue dye, 98, 102
Coptotermes, 17, 179, 180, 233, 254
Corning, Peter, 24
Coronado National Forest, 7, 218
Costa Rica, 4, 85, 89
Couzin, Iain, 152, 273n
Crichton, Michael, 161
Crick, Francis, 98
CRISPRs, 103–104
Crocosphaera, 114
Crohn's disease, 116
Cuban Missile Crisis, 226
Curtis, Tom P., 177
cyanobacterium, 114
cyber security, 222–226
Cyclone Tracy, 174

DARPA, 62n, 214, 219
Darwin (Australia), 174, 175
Darwin, Charles, 45, 51, 77, 111, 174,
    215; social insects and natural
    selection theory of, 22–23, 25, 215

Davidoff, Monte, 142
Dawkins, Richard, 25, 217
"Death of an Order: A
    Comprehensive Molecular
    Phylogenic Study Confirms That
    Termites Are Eusocial
    Cockroaches" (Inward et al.), 6
Defense Department, U.S., 214, 220,
    224
Descartes, René, 23, 68–69, 137, 254
Dhamarrandji, Valerie Milminyina,
    231–34, 240–41
dinosaurs, 7, 77, 91, 176
DNA, 5, 24, 85, 88, 93, 123, 216, 224,
    266n; metagenomic databases of,
    209, 269n; nucleotides of, 96, 103;
    sequencing of, 6, 268n; symbiont,
    185; synthesized, 103–105, 209–11,
    215–16; of viruses, 94
Dominican Republic, 176
Dornhaus, Anna, 151, 153
drones, 146–47, 218–22, 225, 227, 282n
Dubai, 32
Dunlop, Ian, 232, 236
DuPont, 215
Dupré, John, 111

E. coli, 103, 104, 119, 122, 208, 210,
    212–16, 253
Earthwatch, 260n
ecology, 9, 168, 174, 195, 209, 253,
    279n; evolutionary, 124, 173;
    microbial, 177; problem of scale
    in, 206; soil, 194, 203
Egypt, 224, 236
Einem, General Carl von, 221n
Einstein, Albert, 125
electron microscopy, 64, 85

Ellis, John, 77–78, 265*n*, 272*n*
Ellroy, James, 27
Emeryville (California), 99
Empire State Building, 31
Endy, Drew, 270*n*
Energy, U.S. Department of, 3; *see also* Joint BioEnergy Institute
Energy Biosciences Institute (EBI), 85, 86
Engelbrektson, Anna, 10, 12, 86, 91–94, 98, 180
England, 176
Engle, Michael S., 257*n*
entomology, 4, 78, 91–94, 155, 178, 248, 269*n*; mathematics in, 36, 198; at Namibia National Museum, 33–34; of superorganisms, 22; of termite evolution, 6–7
enzymes, 12, 85, 87, 100, 103, 111
ethanol, 91, 100
Ethiopia, 176
eucalyptus trees, 178, 181–82, 190, 192–93, 232, 236
eugenics, 45
European Union, 225
eusociality, 7, 53
euthanasia, 45
Evans, Theo, 20
evolution, 66, 118–23, 139, 153, 173, 179–85, 206, 227, 257*n*; altruism in, 110–11, 163; of antibiotic-resistant bacteria, 177–78; biologists' views on, 42–43, 118, 119, 123–24; of cockroaches into proto-termites, 6–7, 174; of eucalyptus trees, 182–83; of human brain, 215; of microbes, 119–20; of mounds, 34–35, 55–56, 180–81; phylogenic trees in, 93; of social insects, 22–23, 28–29, 45; successes and failures of, 52; symbiotic, 90–91, 109–10, 115, 179, 184–85, 253; technology versus, 215 (*see also* synthetic biology)
extermination, racial, 59, 76–77
extinction, 77

fascism, 45
fairy circles, 196, 204–205
fecal transplants, 116
feedback, 48, 57, 67, 122, 197, 200–201, 204
fertilizer, 124–26, 187, 193, 221, 227, 239
Florida, 254; University of, 155
fluorescein dye studies, 33–34, 38, 58
Formosan termites, 17
Fort Huachuca, 218
Franklin, Rosalind, 98
frequency, 38, 150
Freud, Sigmund, 19, 23, 219
fungi, 6, 58–60, 104, 192–94, 197, 201, 266*n*; in mounds, 16, 21–22, 30, 37–38, 40–42, 52, 56, 58, 87, 157, 205, 248, 251

Gammage, Bill, 275*n*
García Martín, Héctor, 116–24, 128, 154, 207–10, 212–13, 215–17
Garma Festival of Traditional Cultures, 230, 239
gasoline, 4, 7, 83, 100, 125, 207; substitutes for, *see* biofuels
Gates, Bill, 142, 214, 215, 220

genetically modified organisms
(GMOs), 12, 104, 166, 215
genetics, 10–12, 51, 93–97, 106,
179–80, 251, 253; of altruism, 36;
brain, 57; microbial, 3–5, 84–86,
88–90; natural experiments in,
42, 50; palindromic structures in,
103–104; social influence over,
153–54; *see also* DNA
Geneva Conventions, 223, 226
genocide, 59, 77, 221
Genomatica, 214
Germany, 15, 59, 76–77, 125, 176,
191, 221*n*, 255
Gilman, Charlotte Perkins, 45
Glass Bees, The (Jünger), vii, 146,
164, 221, 272*n*
Gnathamitermes perplexus, 92
Goldenfeld, Nigel, 123
Google Earth, 197
Gordon, Deborah, 18, 46
Gove Peninsula (Australia), 190–94,
230, 242, 277*n*
GPS, 180, 219, 222
Grassé, Pierre Paul, 48, 156
grassoline, 4, 83–85, 97, 125, 187, 207,
211–16, 242, 266*n*
Great Barrier Reef, 177, 188
Greece, ancient, 155
Green, Ben, 156
Greengenes database, 177
Grimaldi, David, 287*n*
Guangdong, 10
guano, 124

Haber, Fritz, 125, 221
Haber-Bosch process, 125
Hagerott, Mark, 222–25

Hague, The, 225
Hague Convention (1907), 226
Haldane, J.B.S., 35–36, 198
Hamilton, William D., 36, 198
Harvard University, 60–61, 65, 103,
142, 195, 197, 198, 282*n*; *see also*
Wyss Institute
Hayward Fault, 127–28
He, Shaomei, 86, 91–94, 101–104,
106–108, 110, 112–14, 153, 209
Herero people, 59, 76–77, 221
Herzog, Werner, 277*n*
Himbe people, 59
Hinz, Dieter, 189–95, 229, 230, 233,
235, 236, 239
Hölldobler, Bert, 18, 25
Holocaust, 46
Homeland Security, U.S.
Department of, 7–8
homeostasis, 51–52, 57, 250
Hoover, Herbert, 23*n*
Hoover Dam, 4, 11, 242
horticulture, 189, 191
Hugenholtz, Phil, 3, 85–86, 88–89,
102, 115–18, 122, 168, 188, 208,
229; at Joint Genome Institute,
85, 96–98, 103, 107–10, 121;
"playing jazz" improvisational
science approach of, 90–91,
113–14, 120, 153, 155, 176, 211;
Rosetta stone question of, 112,
115–16, 179, 207; on termite
safaris, 3, 84, 91–95, 174–81, 218,
242; at UC Berkeley, 89
Hurlbut, Benjamin, 228
Hurley, Andrew W., 277*n*
Huxley, Aldous, 23
hydrocyanic acid gas, 46
hydrogen, 87, 111

Illinois, University of, 85, 119
India, 10, 72, 177, 214, 254, 286n
interdependence, tropical pyramid
    of, 10
Internet, 116, 219, 265n
Isoptera, 6
Ivanova, Natalia, 107, 176, 208

Jacklyn, Peter, 180
Japan, 41n, 175
Jasanoff, Sheila, 227, 282n
JFK airport, 40, 56
Jobs, Steve, 146
Johannesburg (South Africa), 15
Joint BioEnergy Institute (JBEI), 85,
    99–100, 105, 117–19, 122–24,
    207–17, 219, 281n
Joint Genome Institute (JGI), 3–4,
    85, 96–98, 105, 107–10, 121, 268n
Journal of Responsible Innovation,
    239
JP-10, 214
Judaism, 221n
Jünger, Ernst, vii, 146, 272n

KARMA programming language,
    226
Keasling, Jay, 100–102, 105, 124,
    209–10, 215, 217; malaria drug
    artemisinin synthesized by, 104,
    214
Keeling, Patrick, 114, 184–85, 206
Kenya, 196, 200–203
Khan, Imran, 146
Kilobots, 142–44, 224
King, Hunter, 286n
Kipling, Rudyard, 35

Kropotkin, Pyotr Alexeyevich, 45
Kruger, Paul, 19
Kuhn, Thomas S., 266n

Las Vegas (Nevada), 95, 178
Lawrence Berkeley National
    Laboratory (LBNL), 85
LAWS, 225–26
Laws of War, 223
Leidy, Joseph, 88
Lenz, Michael, 20
Leonardo da Vinci, 141
Lessing, Doris, 258–59n
lethal autonomous weapons
    systems (LAWS), 225–26
Levin, Simon, 206
"Life Is Physics: Evolution as a
    Collective Phenomenon Far from
    Equilibrium" (Woese and
    Goldenfeld), 123
Life of the White Ant, The
    (Maeterlinck), 24, 252
Light, Sol Felty, 91n
Light Detection and Ranging
    (LIDAR), 197
London, 25
Lost in the Dollhouse, Self-Portrait
    of a Modern Woman (Nagpal), 164
Louisiana, 24, 104
Lüscher, Martin, 30, 48
Ly, Linda, 183–84

machine guns, 76–77, 146, 221, 223,
    226–27, 265n
Macondo oil spill, 104
Macrotermes, 16, 17, 21–22, 29, 39–42,
    87, 196; species of, 50, 156

Mad Max films, 186–87
Maeterlinck, Maurice, 24, 252
Magnetic Island (Australia), 187–89
Maine, 128–30, 135, 149, 157, 254
malaria, 103, 104, 168, 214, 220, 242
Marais, Eugène, 18–25, 28, 34, 52, 69, 79, 224, 250; Maeterlinck accused of plagiarism by, 24, 252; new alphabet concept of, 22, 202; panicking mound described by, 129–30; superorganism musings of, 22–23, 25, 110
Marais, Eugene, 34–35, 39, 43–44, 58, 69, 131; Scott's friendship with, 260n
Margulis, Lynn, 182–83
Marika, Milirrpum, 237
Marika, Wanyubi, 237–38
Marris, Claire, 274n
Mars, 16, 26, 31, 59, 63, 148, 161, 247, 249
Massachusetts, 12, 136–56
Massachusetts Institute of Technology (MIT), 61, 103, 227, 281n
mass spectrometry, 103, 106
Mastotermes darwiniensis, 174–76, 188, 254
mathematics, 28–29, 48, 67, 119, 142–43, 235, 285n; applied, 156; of fairy circles, 195–202, 205; see also biology, mathematical
Matsuura, Kenji, 20
McMahon, Katherine D., 122
metabolisms, 11, 100, 112, 117–20, 129, 208–12, 216
metagenomics, 109–15, 120

methane, 42, 87, 92, 266n
Mexico, 5, 7, 8, 176, 255; Gulf of, 104, 124
microbiology, 88, 93, 114, 182–83, 265n, 269n
Microcerotermes, 179
Microsoft, 142
Middle East, 76
Mitchell, Melanie, 264n
mitochondria, 183, 185
Mixotricha paradoxa, 174, 182–85, 253
Moffett, Mark, 18
Monsanto, 166, 169, 274n
Monte Carlo simulations, 213
Montgomery, David R., 277n
Mott, Cy, 271n
Mughonghora, Vincent, 58
MultiFuel Biofuels (MFB), 213–14
Munungurr, Daymbalipu, 236
muskrats, 128, 130, 271n

Nabalco, 190–91
Nagpal, Radhika, 61–62, 120, 157, 158, 207, 208, 227, 249, 270n; in Namibia, 60–61, 65–67, 70–75, 86, 140; NIH grant of Justin and, 154, 156; at Wyss Institute, 135–36, 140–42, 162–68, 226, 274n
Nalepa, Christine, 20
Namibia, 12, 15–17, 24–25, 84, 175, 186–87, 204; history of, 59, 76–78, 221; National Museum of, 33; NIH funding of termite behavior research in, 154, 156; robotics team in, 60–75, 78–79, 140, 148, 181; termite mounds in, 30, 87, 180, 196, 247–52

nanobots, 161, 224

Napp, Nils, 63–66, 71, 78

Nasutitermes, 4, 17, 99, 180

National Geographic, 15

National Institutes of Health (NIH), 154, 156

National Science Foundation (NSF), 61, 198, 214

natural selection theory, 22–23, 25, 215

Nazis, 46

neurons, 46, 54, 168, 250, 264n

Nevada, 5, 83–86, 91–95

New Orleans, 17

New South Wales (Australia), 186–87

New Yorker, The, 184

Nhulunbuy (Australia), 231–34

Niger River, 77

nitrogen, 87, 111–12, 114–15, 118, 121, 195, 197, 205; fertilizer, 124–26, 187, 193, 221, 227, 239; fixed, 92, 108, 111, 114

Nobel Prize, 24, 99, 252

norovirus, 98

North American termites, 4

Nowak, Martin, 36, 198

nucleotides, 11, 96, 103, 211, 266n

Obama, Barack, 106, 227

Ocko, Sam, 286n

O'Donoghue, P. J. (P.O.D.), 182–83

Odontotermes obesus, 286n

O'Malley, Maureen, 111, 269n

Omatjenne government research farm (Namibia), 26, 260n

Omdurman (Sudan), 76

osteoclasts, 129

Otjiwarango (Namibia), 15, 17, 26, 77, 238

Pacific Northwest, 205

Pakistan, 146, 219

Pangea, 6

Papua New Guinea, 255

Pati, Amrita, 107

PCs, 142

Persian Gulf, 222

Peru, 5, 124

Petersen, Kirstin, 66, 69, 71, 78, 142, 155, 207; TERMES robot work of, 64–65, 135–40, 157–60; videos of termite experiments by, 75, 148–53, 253

Pfams, 109

pheromones, 19–21, 56–57, 75, 153, 253; cement, 48–50, 78, 155–56

photosynthesis, 114, 183

phylogenetic trees, 93, 108–10, 113, 266n

physics, 11, 12, 29, 74, 154, 200, 265–66n; biology and, 116–24, 216–17; condensed matter, 119, 123, 209; in termite research projects, 61, 65, 67, 86, 248–49, 286n

physiology, 12, 29, 30, 51, 65

Pine Barrens (New Jersey), 203

Pinker, Steven, 25

Pinshow, Berry, 65–66, 70–71, 78, 248–49

poison gases, 78, 146, 221n, 226

polymerase chain reaction (PCR), 268n

Predator drones, 218

Prey (Crichton), 161
Princeton University, 195–200, 203–206
Pringle, Rob, 195–99, 203–204, 207, 253, 279n
"Programmable Second Skin to Re-educate Injured Nervous Systems" (NSF grant project), 61
propanediol, 215
P3, 87, 98
protists, 87–88, 110, 173–74, 182–85, 248, 253
protozoa, 182–83

Q64B54, 112
Queensland, University of, 115, 174

Rabinow, Paul, 268n
racism, 44–45, 77
Reade, William Winwood, 77
Reticulitermes, 17
Rhine River, 77
Richardson, Sarah, 211
Rio Tinto, 230, 235, 237, 238
Rirratjingu clan, 237
RNA, 5, 93, 94, 209, 212, 266n
RoboBees, 61–63, 141, 162, 166–69, 219–22, 226
"RoboBees: A Convergence of Body, Brain, and Colony" (NSF grant project), 61
RoboTermite, 63
robots, 12, 60, 70–76, 86, 164–65, 199, 264n, 270n; bee, *see* RoboBees; insights on limitations of, 148–49, 152–54, 163, 253; for

lab work, 98, 210; on Mars, 59, 247; swarming, 143–44, 162, 165; termite, 52, 54, 63–64, 66, 68, 83, 157, 173, 181, 216, 251 (*see also* TERMES); untrustworthiness of, 145–46, 227
Rodgers, Diane, 45n, 261n
Roman Catholic Church, 68
Romania, 198
Roosevelt, Theodore, 226
Roosth, Sophia, 281n
Rosetta stone question, 90, 97, 112, 115–16, 179, 207
Royal Society, 44, 261n
Rubenstein, Mike, 142–44
Russell, Stuart, 229
Rutgers University, 222
Rutherford, Adam, 281n

Sagan, Dorion, 183n
Saha, Krishanu, 228
Saint-Germain-en-Laye (France), Royal Gardens in, 68–69
Sanofi pharmaceuticals, 105, 214
San people, 38
Santa Fe Institute, 264n
Savage Africa (Reade), 77
Scheffer, Marten, 279n
Scheffrahn, Rudi, 4–12, 91, 92, 108
Schivelbusch, Wolfgang, 265n
Scientific American, 164
Scotland, 200, 204
sewage, 88, 93, 113, 121–22
Shakespeare, William, 113
"Shalom" effect, 156
Sheffer, Efrat, 200–203
Shelley, Mary, 161
Sierra Leone, 194

Sikh Golden Temple (Amritsar), 72
Singer, Peter W., 222
Ski Beach (Australia), 240–41
Sleigh, Charlotte, 23n, 261n
slime mold, 61, 199, 263n
Smeathman, Henry, 44, 261n
Smithsonian Institution, 19
Soar, Rupert, 30–31, 37–38, 55–56,
    58, 65, 248–49
Social History of the Machine Gun,
    The (Ellis), 77–78, 265n, 272n
social insects, 23–24, 44–46, 48,
    151, 163, 229, 253; evolution of,
    22–23, 29, 35, 45; see also ants,
    bees
Soo, Rochelle, 188
Soul of the White Ant, The (Marais),
    18–25, 129, 217, 252, 258n
South Africa, 15, 17–19, 24, 29, 32,
    59, 205
South America, 5, 6, 254–55
South West Africa, see Namibia
South West Africa People's
    Organization (SWAPO), 59
Spain, Alister, 187, 189
Spanish, colonization of Central
    and South America by, 5
speculative collaborations, 16
spirochetes, 85, 97, 109–10, 182, 184
statistical physics, 200
STDs, 114
Steig, William, 184
Stephanonympha, 184
stigmergy, 48–50, 54, 66, 156–58,
    185, 253, 263n
string theory, 154
Stubbs, Will, 235
Sudan, 76
Sumatra, 255

Superorganism, The (Hölldobler and
    Wilson), 25
superorganisms, 9, 22–25, 36, 84,
    120, 173, 185, 248; coevolution of,
    91, 111, 115; Marais's musings on,
    22–23, 25, 110, 252; on Mars, 59;
    sewage as, 122; technological, 143,
    219
Sutherland, Jean L., 174
Swammerdam, Jan, 44
swarming technology, 62, 66, 136,
    142, 161–62, 168–69; military uses
    of, 219–22, 225, 227
symbiogenesis, 183–85, 197
symbiosis, 22, 40–41, 94–95, 193,
    182–85, 248; of termite gut
    microbes, 9, 84, 88, 90–91, 110, 115
synthetic biology, 12, 100–106, 161,
    208–17, 253, 270n, 281n; potential
    benefits and risks of, 226–27, 242;
    products created with, 122–24,
    187, 214–15 (see also genetically
    modified organisms)
Synthia, 105, 106, 227

Tahrir Square (Cairo), 224
Talen, Billy (Reverend Billy),
    165–69, 219, 274n
Tarnita, Corina, 195–205, 207, 217,
    249, 253, 279n
TERMES robots, 64, 135–40,
    142–43, 145, 157–62
Terminix, 10
Termitomyces fungus, 41, 58
Théraulaz, Guy, 136, 263n
thermodynamics, 74, 119, 121
This Is My Thinking (film), 236, 284n
3-D models, 64, 156

3-D printing, 138–39, 167, 220

Tibet, 76, 265n

Toowoomba (Australia), 189–90, 194

Towers of Hanoi logic game, 138

Townsville (Australia), 188

Trichomonas, 114

Trichonympha, 88, 110, 253

trophallaxis, 23n, 151, 188

tropics, pyramid of interdependence of, 10

Turing, Alan, 199, 200, 279n

Turner, J. Scott, 15–17, 25–31, 60, 65, 84, 149, 248, 253; Eugene Marais and, 33–35, 39, 43, 58, 260n; evolutionary views of, 28–29, 66, 90; lab work of, 27, 48–50, 53–58, 63, 207; mounds studied by, 27–30, 37–39, 41–43, 129, 180; on termite cognition, 51–52, 56–57, 120, 181, 219, 250–51; termite individuality recognized by, 154–56

Twitter, 48

United Kingdom, 226

United Nations (UN), 225

United States, 15, 22, 41n, 57, 72, 146, 226, 255; airplane development in, 168; green technology initiative in, 211–12 (see also specific research programs); military technology of, 7, 214, 219–26; wastewater treatment plants in, 93; see also specific states and municipalities

UNIX, 108

utopias, 44–45, 163, 214, 221, 243, 251

Venter, Craig, 105–106, 124, 216

Versailles, Treaty of, 125

Vietnam, 214

viruses, 94, 98, 103, 116, 219

Von Trotha, Lothar, 76, 220–21

Waiting for Godot (Beckett), 74

Walnut Creek (California), 96

Wanambi, Wukun, 235

Warnecke, Falk, 89–90, 94, 95, 97–98

Washington, D.C., 124

wastewater treatment plants, 93, 113, 121–22

Waterberg Plateau (Namibia), 77

Watson, James, 98

weaverbirds, 47

Werfel, Justin, 67–68, 73–75, 143, 157, 160, 161; National Institutes of Health grant of, 154, 156; TERMES robots of, 135–38, 140; tunnel experiment of, 78–79; Waiting for Carnot approach of, 73–74, 119

Wheeler, William, 22–23, 45, 54, 116; as originator of term "superorganisms," 22

whegs, 136–38

Where the Green Ants Dream (film), 277n

white supremacy, 77

Williams, Nancy M., 190–91

Wilson, E. O., 18, 24, 25, 110, 198

Windhoek (Namibia), 15, 16, 59

Woese, Carl, 123, 216, 265–66n

Wood, Rob, 61

Woody Beach (Australia), 240

World War I, 23n, 59, 79, 125, 146, 221n, 223, 226

World War II, 35, 175

Wyoming, 176
Wyss Institute, 61, 135–45, 165, 226

XCell SureLock, 102
X-rays, 266n

Yemen, 219
Yirrkala (Australia), 229, 231, 234,
    237, 240, 242, 284n

Yolngu people, 190–92, 229, 233,
    236, 277n
Yothu Yindi (band), 230
Ypres (Belgium), 221n
Yunupingu, Mungurrawuy,
    241
Yunupingu, Yalmay, 230

zoology, 45
Zyklon B, 221n

304

A NOTE ABOUT THE AUTHOR

Lisa Margonelli is the author of *Oil on the Brain: Petroleum's Long, Strange Trip to Your Tank*, which won a Northern California Book Award and was named a notable book of the year by the American Library Association. She is deputy editor at *Zócalo Public Square*, an Arizona State University magazine of ideas. From 2006 to 2012, she was a fellow at the New America Foundation. She has written for *The Atlantic, Wired, Scientific American, The New York Times*, and other publications. She lives in Maine.

A NOTE ABOUT THE ILLUSTRATOR

Thomas Shahan is an artist and macro photographer from Oklahoma with a passion for entomology and relief print-making. When he's not chasing jumping spiders in his backyard, he's carving linoleum block prints.